Springer-Verlag Berlin Heidelberg GmbH

Paul K.Y. Wong   William S. Lynn (Eds.)

Neuroimmunodegeneration

Springer

Paul K.Y. Wong

The University of Texas
M.D. Anderson Cancer Center
Science Park-Research Division
Smithville, Texas, U.S.A.

William S. Lynn

The University of Texas
M.D. Anderson Cancer Center
Science Park-Research Division
Smithville, Texas, U.S.A.

ISBN 978-3-662-12581-6

Library of Congress Cataloging-in-Publication Data

Neuroimmunodegeneration / [edited by] Paul K. Y. Wong, William S. Lynn.
    p. cm.-- (Biotechnology intelligence unit)
Includes bibliographical references and index.
    ISBN 978-3-662-12581-6   ISBN 978-3-662-12579-3 (eBook)
    DOI 10.1007/978-3-662-12579-3

    1.   Nervous system—Degeneration—Immunological aspects.
2. Neuroimmunology. I. Wong, Paul K. Y.II. Lynn, William S.III. Series.
    [DNLM: 1. Central Nervous System Diseases—physiopathology.
2. Central Nervous System Diseases—immunology. 3. Nerve Degeneration—
physiopathology. 4. Immune System—pathology.   WL 300 N49381553 1998]
RC365.N477   1998
616.8'0479—dc21
DNLM/DLC
for Library of Congress   98-23047
                                                                           CIP

© Springer-Verlag Berlin Heidelberg 1998
Originally published by Springer-Verlag Berlin Heidelberg New York in 1998

Typesetting: R.G. Landes Company, Georgetown, TX, U.S.A.

SPIN 10685048   31/3111 - 5 4 3 2 1 0 - Printed on acid-free paper

DEDICATION

To Pick-Hoong and Mary

PREFACE

Neuroimmunodegeneration (NID) is the term applied to diseases in which immune and neural cells die prematurely. Because the central nervous system (CNS) has its own immune/support system composed primarily of astrocytes, those diseases in which neuronal loss is associated with impairment of CNS immune/support cells are also considered NID syndromes. Studies of some NID syndromes suggest that the primary cause of neuronal and immune cell death is perturbation of signaling pathways that control apoptosis, differentiation and proliferation. Such perturbations can be initiated by viral infection or by defective genes. Examples of NID syndromes are ataxia telangiectasia, Alzheimer's disease, Parkinson's disease, Down syndrome, and several prion and retroviral diseases. These NIDs are usually associated with aggregation and accumulation of cytotoxic or inflammagenic fibrillary proteins. These fibrillary proteins, which are produced by astrocytes and neurons, include phosphorylated neurofilaments, amyloid precursor proteins, presenilins, apolipoprotein E, prion proteins, and retrovirus envelope proteins. In most of these fibrillary protein-associated diseases, the CNS immune/support system is locally upregulated without alteration of the peripheral immune system. However, in syndromes in which the support cells in the CNS are intensely mitogenically activated, as in transgenic mice whose astrocytes overexpress inflammatory cytokines, a massive influx of activated peripheral immune cells through leaky blood brain barrier may also contribute to neuronal death.

Chapter 1 presents the intrinsic relationship between the nervous and immune systems. Not only are both of these systems derived from the neural crest, but many of the same cytokine signals are used to regulate prenatal development and postnatal growth of the two systems. Identification of the function of these cytokine signals during pre- and postnatal development should provide insight into the pathogenic mechanisms involved in NIDs.

Chapter 2 provides a definition of the NID syndromes and presents several examples of human and animal models for NIDs. All of these NID syndromes and models involve neurodegeneration associated with impairment or dysfunction of the CNS immune/support system. However, in some cases, such as ataxia-telangiectasia and retrovirus-induced NID syndromes, neurodegeneration is also associated with dysfunctions of the peripheral immune system.

Although mechanisms involving various NID syndromes are still sketchy, Chapter 3 attempts to identify the potential mechanisms involved in cell losses in NID syndromes. The interactions and functions of the major organelles that control cell fate, the effects of accumulation of neurotoxic fibrillary proteins, and the role of the CNS immune/support cells in neuronal loss are reviewed.

Chapters 4 and 5 present two murine retrovirus animal models of NID that produce different effects in the immune and nervous systems. These chapters illustrate the importance of animal models in the study of the pathogenic mechanisms of NID and their potential in developing therapies for these diseases.

Recent advances in transgenic technology have provided new tools and approaches for investigation of specific features involved in pathogenesis that could not be effectively studied before. Chapters 6 and 7 describe the generation of transgenic mice targeted to overexpress proinflammatory cytokines in the CNS to elucidate the role and potential mechanisms of these cytokines in neuronal cell death.

Although still in its infancy, the field of neuroimmune interactions is growing tremendously. We believe that this book on NID syndromes, which attempts to deal with premature cell losses in both the nervous and immune (peripheral and central) systems, may help us to understand some of the pathogenic mechanisms involved in the various interactions between the central and peripheral immune/defense systems which modulate the pathways controlling the fate of neurons, astrocytes, and T cells.

Paul K.Y. Wong, William S. Lynn
U.T.M.D. Anderson Cancer Center

CONTENTS

EDITORS

Paul K.Y. Wong, Ph.D.
The University of Texas
M. D. Anderson Cancer Center
Science Park-Research Division
Smithville, Texas, U.S.A.
Chapter 2, 3, 4

William S. Lynn, M.D.
The University of Texas
M. D. Anderson Cancer Center
Science Park-Research Division
Smithville, Texas, U.S.A.
Chapters 2, 3, 4

CONTRIBUTORS

Katerina Akassoglou, Ph.D.
Hellenic Pasteur Institute
Department of Molecular Genetics
Athens, Hellas
Chapter 7

Anthony S. Basile, Ph.D.
National Institutes of Health
Laboratory of Neuroscience
NIDDK
Bethesda, Maryland, U.S.A.
Chapter 5

Iain L. Campbell, Ph.D.
The Scripps Research Institute
Department of Neuropharmacology
La Jolla, California, U.S.A.
Chapter 6

Wonkyu Choe, Ph.D.
The University of Texas
M. D. Anderson Cancer Center
Science Park-Research Division
Smithville, Texas, U.S.A.
Chapter 4

Michael Graham Espey, Ph.D.
National Institutes of Health
Laboratory of Neuroscience
NIDDK
Bethesda, Maryland, U.S.A.
Chapter 5

George Kassiotis, BSc
Department of Molecular Genetics
Hellenic Pasteur Institute
Athens, Hellas
Chapter 7

John A. Kessler, M.D.
Albert Einstein College of Medicine
Rose F. Kennedy Center
    for Research in Mental Retardation
    and Human Development
Bronx, New York, U.S.A.
Chapter 1

George Kollias, PhD
Hellenic Pasteur Institute
Department of Molecular Genetics
Athens, Hellas
Chapter 7

Yelena Kustova, Ph.D.
National Institutes of Health
Laboratory of Neuroscience
NIDDK
Bethesda, Maryland, U.S.A.
Chapter 5

Y.C. Lin, Ph.D.
The University of Texas
M. D. Anderson Cancer Center
Science Park-Research Division
Smithville, Texas, U.S.A.
Chapter 4

Eliezer Masliah, M.D.
University of California, SanDiego
Departments of Neuroscience
   and Pathology
San Diego, California, U.S.A.
Chapter 6

Mark F. Mehler, M.D.
Albert Einstein College of Medicine
Rose F. Kennedy Center
   for Research in Mental
   Retardation and Human
   Development
Bronx, New York, U.S.A.
Chapter 1

Axel Pagenstecher, M.D.
University of Freiburg
   and Breisacherstr.
Department of Neuropathology,
   Pathology
Freiburg, Germany
Chapter 6

Lesley Probert, Ph.D.
Hellenic Pasteur Institute
Department of Molecular Genetics
Athens, Hellas
Chapter 7

Yoshitatsu Sei, M.D., Ph.D.
Uniformed Services
   University of the Health Sciences
Department of Anesthesiology
Bethesda, Maryland, U.S.A.
Chapter 5

Anna K. Stalder, Ph.D.
The Scripps Research Institute
Department of Neuropharmacology
La Jolla, California, U.S.A.
Chapter 6

P.H. Yuen, Ph.D.
The University of Texas
M. D. Anderson Cancer Center
Science Park-Research Division
Smithville, Texas, U.S.A.
Chapter 4

CHAPTER 1

Hematolymphopoietic and Associated Cytokines in Neural Development

Mark F. Mehler and John A. Kessler

It has become increasingly apparent that many of the same factors regulate development of both the nervous and hematolymphoid systems. Members of two cytokine superfamilies, the hemopoietins and the transforming growth factor βs (TGFβs), and the more recently characterized glial cell-derived neurotrophic factor (GDNF) family, mediate a complementary range of developmental events in the nervous system that frequently exceed those mediated by the classic neurotrophins (Table 1.1). These cytokine families, which are active in hematolymphoid development, are also involved in neurulation, dorsoventral patterning of the neural tube, and in the progressive evolution of the mammalian central and peripheral nervous systems. Recent studies have begun to characterize the detailed receptor subunit organization and the intracellular signaling pathways utilized by these growth factor families, the specific environmental contexts in which they act and the range of cellular processes by which they orchestrate the progressive sculpting of the developing nervous system. In addition, several experimental investigations have presented provocative evidence suggesting a role for these cytokines in modulating a range of developmental events through bidirectional communications between the hematolymphoid and neural systems.[1-4] These diverse cytokines exhibit significant functional redundance and pleiotropy, especially during brain development; thus, targeted knockouts of cytokine signaling components generally demonstrate few central nervous system deficits, although there are often significant developmental abnormalities in the peripheral nervous system.

Neuroimmunodegeneration, edited by Paul K.Y. Wong and William S. Lynn.
© 1998 Springer-Verlag and R.G. Landes Company.

Table 1.1. Cytokines active during neural development

Hematopoietic Factors:

  I. γ_c Receptor Subunit Factors
    A. Interleukins (IL-s) 2,4,7,9,15
    B. Variants: IL-4, -12, -13

  II. β_c Receptor Subunit Factors
    A. IL-3, -5, Granulocyte-Macrophage (GM)-Colony Stimulating Factor (CSF)

  III. GP130-Related Subunit Factors
    A. Leukemia Inhibitory Factor Receptor β (LIFRβ)/gp130 Heterodimeric Subunit-Mediated
      1. Cilary Neurotrophic Factor (CNTF), LIF, Oncostatin-M (OM), Cardiotrophin-1 (CT-1)
    B. GP130 Homodimeric Subunit-Mediated
      1. IL-6, -11
      2. Variants: Granulocyte (G)-CSF, Leptin

  IV. Interferon (IFN)-Related Factors
    1. IFNα/β, γ, IL-10

  V. Tumor Necrosis Factor (TNF) Family
    1. TNFα, Fas-L

  VI. Chemokines
    1. IL-8

  VII. Erythropoietin (EPO)/Thrombopoietin (TPO) Family
    1. EPO, TPO

  VIII. Additional Interleukins
    1. IL-1

  IX. Tyrosine Kinase Receptor-Mediated
    1. Stem Cell Factor (SCF), CSF1, FLT-3L

Transforming Growth Factor β (TGFβ) Superfamily Factors:
  A. TGFβ1-3
  B. Activin A
  C. Bone Morphogenetic Protein (BMP) Subclass
    1. BMPs 2,4,5,6,7,9,12,13
    2. Growth/Differentiation Factor (GDF) 5
    3. Dorsalin-1

Glial-Cell-Derived Neurotrophic (GDNF) Factor Family:
  A. GDNF, Neurturin-1

Neurulation and Dorsoventral Patterning

In the dorsal blastopore lip, the vertebrate organizer contains a number of bone morphogenetic protein (BMP)-related factors and regulatory proteins that are involved in neurulation and dorsoventral patterning.[5] The BMPs, with more than 20 members, represent the largest and most rapidly expanding subclass of the transforming growth factor β (TGFβ) superfamily of ligands that transduce intracellular signals through serine-threonine kinase receptor subunits.[6,7] BMPs mediate a diverse array of developmental processes, including cellular survival, proliferation, morphogenesis, lineage commitment, inhibition of alternate lineages, differentiation and apoptosis (Table 1.2).[5,7-12] Both BMP4 and BMP7 are expressed in the ectoderm, and are epidermal inducers associated with concurrent inhibition of neurulation.[8,13] By contrast, activin inhibits neurulation by induction of mesoderm rather than epidermis.[13] Dominant-negative forms of the type I BMP receptor promote neurulation; in vitro this cellular action is blocked by application of BMP4.[14,15] Similar phenotypes are apparent with dominant-negative forms of BMP4 and 7.[14,15] Within the Spemann organizer, chordin, noggin and follistatin are structurally-distinct proteins that antagonize BMP signaling and establish a series of complex gradients necessary for later morphogenesis.[16-19] These molecules bind directly to the BMPs and restrict regional expression, block ligand activity and also compete at the level of the BMP receptors.[5,17-19] Follistatin is also expressed in the Spemann organizer and in the notochord and exhibits direct neuralizing activity.[20] Thus, inhibition of BMP signaling appears to be required for the conversion of ectoderm to neuroectoderm. Animal cap cell dissociation and dominant-negative experiments further suggest that cells of the gastrula animal cap are predisposed to form neural tissue in the absence of additional signals, and that epidermal fate is an "induced" state of ectoderm dependent upon BMP signaling.[13] Anterior neurulation results from an inhibition of BMP4 signaling. Analysis of BMP4 homozygous null mutants, however, has shown that excess neural tissue is not induced at the expense of epidermal ectoderm. Therefore, other BMPs may function through an alternate pathway to inhibit neurulation.[5] In more caudal regions, fibroblast growth factor may be required for neural specification, either alone within posterior regions (that later give rise to the spinal cord), or in combination with additional proteins such as noggin that are required in intermediate regions (for subsequent generation of the mid- and hindbrain).[21]

Table 1.2. Cellular actions of BMPs during neural development

Inhibits neurulation during gastrulation

Neuralizing factors (Chordin, Noggin, Follistatin) bind to and differentially inhibit the diffusion and cellular actions of the BMPs

Induces dorsal cell types in the neural tube during dorsoventral patterning (neural crest cells, sensory and commissural neurons)

Antagonizes the ventralizing actions of Sonic Hedgehog in the neural tube: Modulates expression of intermediate cell types

Promotes neuronal lineage commitment during neural crest development

Promotes expression of adrenergic sympathetic neurons from postmigratory neural crest progenitor cells

Potentiates expression of neurotransmitter and neuropeptide genes in sympathetic neurons

In concert with nerve growth factor, promotes neurite outgrowth from sympathetic neurons

Promotes astroglial lineage commitment from multipotent and oligopotent embryonic subventricular zone and postnatal subcortical white matter progenitor species

In concert with LIFβ receptor activation, promotes expression of radial glia and radial glial pathway-dependent astrocytes from embryonic subventricular zone progenitor cells

Potentiates survival and cellular differentiation of embryonic CNS neuronal progenitor cells

Potentiates expression and maturation of embryonic ventral mesencephalic dopaminergic neurons

Potentiates expression of neural-specific gene transcripts and neuronal terminal differentiation of PC12 cells

Promotes regional apoptosis of cultured rhombencephalic neural crest-derived odd-numbered (r3,5) rhombomeres.

BMPs are also important for the dorsoventral patterning of the developing neural tube. Non-neural cells flanking the dorsal neural tube secrete inductive signals that generate neural crest stem cells, sensory neurons and roof plate cells.[5,8] The BMP-related molecule, dorsalin-1, is expressed in the dorsal neural tube and induces appropriate regional cell types in vitro.[5] BMP4 and BMP7 are expressed at appropriate developmental stages in chick non-neural ectoderm, and these cytokines induce dorsal cell types in neural plate explants.[5] As the neural plate gives rise to the developing neural tube, transcripts for BMP4 and BMP7 are downregulated in the epidermis, but transcripts for BMP7 continue to be expressed within forebrain.[5,8,22] Further, significant levels of BMP4 transcripts are present within dorsal neural folds and in the midline of the neural tube as it closes. In murine tissues, however, BMP2 rather than BMP4 is initially expressed in the anterior neural folds.[5,23] After closure of the neural tube and cellular maturation, BMP4 is expressed in the anterior dorsal midline region, BMP6 along the whole neuraxis, and BMPs 5 and 7 in partially overlapping domains of the dorsal midline.[5,8] BMPs and Sonic Hedgehog, a secreted factor produced by the notochord and floorplate that induces ventral cell types,[8,24] appear to be mutually repressive, but exhibit cooperative interactions to induce the elaboration of intermediate cellular species that exhibit distinct phenotypic profiles.[8] In pre-migratory dorsal-derived neural crest cells, BMPs induce expression of Pax1 and Msx1 and the zinc finger transcription factor, Slug. Within the dorsal neural tube, the BMPs exert their developmental actions through both local and long-range signaling. Long-range signals are propagated by specific BMP binding proteins (e.g., noggin, chordin, follistatin) that are developmentally regulated and establish morphogenetic gradients necessary for the differential induction of dorsal cell types.[5,8,17-19] In *Drosophila*, the chordin homologue, Sog, is freely diffusible and is able to selectively bind Dpp (the homologue of BMP2 and BMP4).[25] The zinc metalloprotease homologue, Tolloid, can cause release of Dpp, and these dynamic interactions determine the gradient of Dpp signaling.[5,25] Thus, Dpp acts as a diffusible extracellular morphogen that is responsible for the induction of specific dorsal and intermediate cell types.[5,8] During *Xenopus* and murine neurogenesis, the three distinct BMP binding proteins further refine the morphogenetic gradient established during dorsoventral patterning, through differences in expression patterns, ligand targets and binding affinities, diffusion rates, levels of transcript expression and associated variables.[17-19]

Early CNS Development

Neural stem/progenitor cells derived from the developing neural tube along its entire rostrocaudal axis (e.g., telencephalon to the spinal cord) initially undergo symmetric cell divisions to exponentially expand the number of self-renewing progeny within periventricular generative zones from the spinal cord to the cerebral cortex.[26,27] Early neural mitogens (epidermal growth factor, basic fibroblast growth factor) and related factors orchestrate the activation, proliferation, viability and early lineage decisions of these multipotent progenitors.[26,27] Neuropoietic cytokines, including CNTF, LIF, and OM potentiate commitment to the astroglial lineage from early neural progenitor cells (Fig. 1.1). LIFβ receptor-blocking antibodies effectively attenuate this cellular action in cultures of early embryonic spinal cord neural progenitors.[28] Multipotent progenitors give rise to oligodendroglial progenitors (OLPs) under the influence of nonhemopoietin oligotrophins in late embryonic and early postnatal development in a caudal-rostral gradient.[29] Although CNTF and IL-4 mediate the survival of early rodent OLPs in vitro, long-term survival requires the presence of growth factors from two additional nonhemopoietin subclasses in addition to the neuropoietic cytokines.[30] The need for combinatorial cytokine interactions to potentiate specific CNS developmental processes is a recurring theme in early neural development. Early rat OLP mitogenesis in vitro is promoted by IL-2 and IL-4, and negatively regulated by TGFβ superfamily factors.[29,31-34]

BMPs also induce the dose-dependent elaboration of astrocytes from epidermal growth factor (EGF)-responsive embryonic murine subventricular zone (SVZ)-derived multipotent progenitor species.[7,35,36] These developmental actions are associated with early elaboration of an immature astroglial phenotype, inhibition of cellular proliferation and later terminal differentiation, including increased cellular expression of glial fibrillary acidic protein, GFAP.

Fig. 1.1. (opposite) Diverse roles of hematolymphopoietic and inflammatory cytokines in the development of the central (CNS) and peripheral (PNS) nervous systems. During neural development, these cytokines are involved in multiple aspects of lineage restriction, commitment, progenitor cell proliferation and survival, and neuronal differentiation, including transmitter/receptor phenotypic expression, axodendritic process outgrowth and synaptic regulation. Abbreviations: ACT A, activin A; AS, astrocyte; bFGF, basic fibroblast growth factor; BMP, bone morphogenetic protein; CN, cholinergic neuron; CNTF, ciliary neurotrophic factor; CSF, colony-stimulating factor; CT-1, cardiotrophin-1; EPO, erthropoietin; G-, granulocyte-; GGF, glial growth factor; GM-, granulocyte-macrophage; IFN, interferon; IL-, interleukin; LIF, leukemia inhibitory factor; M, melanocyte; MP, multipotent progenitor; N, neuron; NCSC, neural crest stem cell; NP, neuronal progenitor; NT, neurotrophin; NTSC, neural tube-derived stem cell; OL, oligodendrocyte; OM, oncostatin-M; OP, oligodendroglial progenitor; PN, parasympathetic neuron; RG, radial glia; SAN, sympathoadrenal neuronal progenitor; SC, Schwann cell; SCF, stem cell factor; SM, smooth muscle cell; SN, sensory neuron; SNN, sensory neuroblast; SYM, sympathetic neuron; SYMN, sympathetic neuroblast; TGF, transforming growth factor; TNF, tumor necrosis factor; TPO, thrombopoietin. (Reprinted with permission from Trends in Neurosciences 20:359, 1997, Elsevier Science Ltd.)

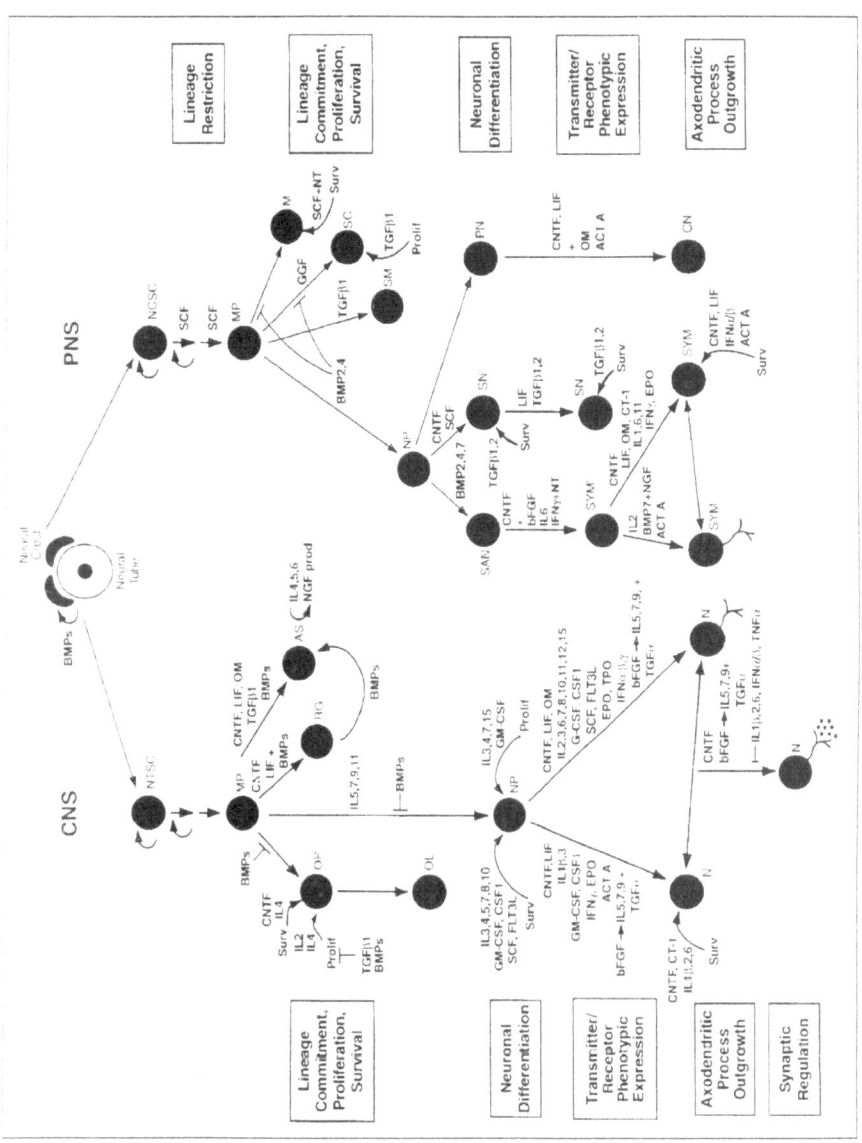

Postnatal subcortical bipotent oligodendroglial-astroglial (O-2A) progenitor species are also responsive to the BMPs. O-2A progenitor cells, in vitro, normally give rise to differentiated oligodendrocytes when propagated in serum-free medium (SFM) and to type II astrocytes when propagated in 10% fetal calf serum.[7,35,36] Following transplantation, O-2A progenitor species selectively give rise to cells of the oligodendrocyte lineage unless placed in a glial-deficient environment. Under these latter pathological conditions, bipotent progenitor species can generate both oligodendrocytes and astrocytes.[37] When plated and propagated in SFM, BMPs promote the dose-dependent elaboration of type II astrocytes associated with premature exit from the cell cycle.[7,35,36] For both SVZ-derived and O-2A progenitor species, these developmental actions are associated with concurrent suppression of alternate lineage elaboration: inhibition of both the neuronal and oligodendroglial lineages (SVZ progenitor cells) or only the oligodendroglial lineage (O-2A progenitor species).[7,38] Although the regulation of gliogenesis during mammalian development is poorly understood, in *Drosophila, glial cells missing (gcm)* regulates lateral glial development with subsequent glial differentiation mediated by the downstream target gene, *pointed*, and neuronal maturation modulated by a splice variant of the downstream gene, *tramtrack*.[39] Thus, it is quite possible that mammalian homologues of the *gcm* gene cascade may be downstream targets of BMP actions during CNS gliogenesis.

In concert with a specific hemopoietin subclass (gp130 heterodimeric cytokines), BMPs also act to potentiate the generation of radial glia from EGF-generated embryonic SVZ multipotent progenitor cells. Radial glia, forerunners of the astroglial lineage, exist within embryonic CNS cortical generative zones during neurogenesis and act as a scaffold to sanction neuronal migration and cellular differentiation.[40] Leukemia inhibitory factor (LIF), a cytokine that signals through gp130 heterodimeric receptors, in combination with BMP2 or basic fibroblast growth factor (bFGF), potentiates the elaboration of radial glia from late embryonic SVZ progenitor cells, with subsequent enhancement (BMP2) or inhibition (bFGF) of the generation of GFAP-immunoreactive astrocytes.[7,41] Detailed clonal analysis further reveals that LIF targets only early multipotent progenitor species, while BMP2 may target both early and more lineage-restricted progenitor species.

CNS Neuronal Lineage Development

Fibroblast growth factor predisposes progenitor cells to undergo neuronal differentiation, but additional signals are necessary to potentiate neurogenesis. IL-5, -7, -9 and -11 have been shown to potentiate the generation of neuroblasts from a conditionally-immortalized cell line derived from embryonic murine hippocampal multipotent progenitor cells.[42] These developmental actions are associated with increased expression of neurofilament proteins, decreased cellular electrotonic coupling and the development of embryonic sodium channels and action potentials, landmarks in the establishment of the early neuronal lineage.[42] After neuroblast lineage commitment, these evolving progenitor species require a specific complement of regionally-expressed cytokines to promote survival and to mediate continued cellular proliferation, prior to undergoing progressive stages of terminal differentiation. Interleukins 3, 4, 5, 7, 8, 10, GM-CSF, CSF1, SCF and FLT3L potentiate cellular survival of cultured embryonic murine and rodent neuroblasts derived from multiple brain regions.[8,12,43-47] These survival effects are region-selective, dose-dependent, mediated by direct factor actions (IL-7), paracrine signaling (CSF1) or both (GM-CSF), linked to actions on axodendritic process-outgrowth (IL-8, -10) or enhanced by the additional cellular effects of complementary cytokines (SCF). Cellular proliferation of these cultured embryonic neuroblast subpopulations is potentiated by IL-3, -4, -7, -15 and GM-CSF.[8,12,44]

Progressive neuronal differentiation and phenotypic maturation require the expression and modulation of neurotransmitters, associated neuromodulators and biosynthetic enzymes within regional neuronal subpopulations. CNTF exerts minimal maturational effects (e.g., expression of tyrosine hydroxylase) on cultured striatal neurons, and always requires cofactors for similar actions on embryonic rat cortical neurons in vitro.[48] CNTF or LIF also potentiates the expression of choline acetyltransferase (ChAT) when coapplied with a neurotrophin to cultured human spinal cord neurons.[49] LIF also enhances the expression of ChAT when applied to cultured rat motor neurons, with additional actions on cellular survival.[50] IL-3 and GM-CSF potentiate the expression of glutamic acid decarboxylase (GAD) and gamma-amino-butyric acid (GABA) in cultured murine septal neurons, while erythropoietin and CSF1 also increase ChAT expression in embryonic murine septal neuroblasts.[51-53] In addition, TGFβ2 and 3 promote the survival of cultured dopaminergic neurons.[54] Specific TGFβs additionally enhance neuronal survival in concert with other cytokines (basic fibroblast growth factor) and modulate levels of neurotrophins produced by target tissues.[54] TGFβ has been shown to regulate homeostasis and to accentuate the expression of the survival gene, *bcl-2* in rat hippocampal neurons. This cytokine can also further protect these cells against oxidative and cytotoxic injury by potentiating calcium buffering and mitochondrial function.[55,56] Exposure of a conditionally-immortalized embryonic hippocampal neural progenitor cell line to IL-5, -7 or -9 in combination with transforming growth factor α (TGFα) enhances the maturation of neuronal voltage-gated sodium channels and the later expression of more phenotypically-mature ligand-gated channels.[42]

An extensive spectrum of hemopoietins also promote additional aspects of terminal neuronal differentiation, including axodendritic process outgrowth and the expression of neurofilament proteins. Cytokines that signal through both gp130 homodimeric (IL-6, -11, G-CSF) and heterodimeric (CNTF, LIF, OM) receptors enhance phenotypic maturation and axodendritic process outgrowth of primary murine and rodent cultured embryonic neuroblasts from multiple CNS regions; greater neurite elaboration is associated with addition of gp130 homodimeric factors.[8,12] A distinct spectrum of neuronal maturational effects on these CNS regional cultured embryonic neuroblasts is also apparent following exposure to interleukins 3, 4, 7-10, 12, 13 and 15.[8,12,44,47] An additional group of hemopoietins (CSF1, SCF, FLT3L, EPO, TPO) further enhances these neuronal maturational indices within complementary neuroblast subpopulations derived from these primary cultured embryonic brain regions; several of these cytokines also enhance neuroblast maturation from cultured embryonic murine and rodent subventricular zone multipotent progenitor cells.[8,12,47] In vitro, interleukin 2 promotes hippocampal neuronal maturation, while IL-3 enhances neurite outgrowth from central cholinergic neurons both in vitro and in vivo.[8,12,51,57] While IFN α/β potentiates cellular differentiation and the expression of mature neurofilament proteins in cultured fetal murine neuroblasts, IFNγ has complementary cellular effects on cultured hippocampal and cortical neuroblasts.[58-60] Further, coapplication of IL-5, -7 or -9 with TGFα following bFGF administration increases expression of mature neurofilament proteins and promotes terminal axodendritic process elaboration in neuroblasts derived from a conditionally-immortalized hippocampal neural progenitor cell line.[42]

During progressive stages of neuronal maturation, a select group of cytokines positively modulates cellular survival. CNTF enhances the survival of chick spinal cord motor neurons in vivo, and both rat and murine hippocampal neuronal subpopulations in vitro, while cardiotrophin-1 (CT-1) potentiates cellular viability of multiple CNS neuronal subpopulations in vitro.[54] In addition, IL-1β, in concert with IL-2, potentiates the survival of cultured spinal cord, basal forebrain and hippocampal neurons, while IL-6 promotes the survival of cultured rat mesencephalic, hypothalamic, catecholaminergic and postnatal septal cholinergic neurons.[54,57,61]

The final stages of neuronal differentiation encompass the processes of synaptogenesis, regulation of mature neuronal electrical excitability and synaptic plasticity. CNTF regulates spontaneous and impulse-associated neurotransmitter release in nascent *Xenopus* motor neurons in vitro.[57] In relation to the conditionally-immortalized neural progenitor cell line, the previously mentioned cytokine permutations accentuate the expression of mature neuronal action potentials that are sensitive to the actions of tetrodotoxin. Several cytokines, including IL-1β, -2, -6, IFN α/β and TNFα, display a dose-dependent inhibition of long-term potentiation of hippocampal neurons in vitro. Further, leptin exerts pleiotropic effects on the regulation of feeding behavior and energy levels; these biologic actions are mediated directly by hypothalamic neurons.[62] Within hematopoietic development, leptin enhances the proliferation and differentiation of progenitor cells, mediated by the nontruncated OB-Rb receptor.[63] Leptin transcripts are expressed by the bone marrow stroma, and leptin receptor mRNA is present within early hematopoietic cells.[64,65]

Early PNS Development

For early peripheral nervous system (PNS) development in the chick, premigratory neural crest stem cells undergo lineage restriction in vitro under the influence of SCF, a hemopoietin that targets multiple stages of the neural crest developmental cycle.[66] Multipotent progenitor cells initially give rise to an extensive range of lineage species, including neuronal progenitors, glial species (Schwann cells), smooth muscle and melanocytes (Fig.1.1). Melanocyte survival is enhanced by SCF when coupled with a neurotrophin.[66] Neural crest-derived lineage commitment is instructive and is directly regulated by neuregulins (Schwann cells: glial growth factor, GGF) or TGFβ subclass factors (neuronal progenitors: BMP2, 4; smooth muscle: TGFβ1).[67-69]

PNS Neuronal Lineage Development

During early stages of neuronal differentiation there is elaboration of sympathoadrenal and sympathetic neuroblasts, embryonic spinal sensory neuroblasts, including dorsal root ganglion neurons, and parasympathetic ciliary neuroblasts.[67] Within the developing PNS, BMP2 or 4 plays an instructive role in the elaboration of neuronal lineage species from neural crest stem cells, in association with the induction of Mash 1, a basic helix-loop-helix transcription factor.[67,68] At saturating doses, the BMP-mediated neuronal instructional signals are dominant over those of the neuregulins, which play a complementary biological role in the induction of the glial lineage (Schwann cells).[68] In addition, BMP4 and BMP7 are expressed within the dorsal aorta during a critical period of postmigratory neural crest-derived sympathetic neuronal differentiation.[70] In the quail system, BMPs 2, 4 and 7, but not BMP6, induce elaboration of adrenergic sympathetic (tyrosine hydroxylase-immunoreactive) and an additional postmitotic (islet-1-reactive) neuronal subset.[71,72] These developmental actions are concentration-dependent, suggesting the presence of a hierarchy of BMP subgroup effects.[71,72] BMPs also exert biological actions at later stages of rat sympathetic neuronal differentiation.[73] Thus, BMP2 and BMP6 induce distinct profiles of neurotransmitter and neuropeptide expression without changing levels of transcripts for tyrosine hydroxylase.[73] Coapplication of BMP7 with nerve growth factor promote selective dendritic outgrowth from sympathetic neurons.[74] Exposure to both CNTF and bFGF induce early differentiation of an immortalized sympathoadrenal (MAH) cell line prior to the occurrence of cellular responsiveness to nerve growth factor (NGF).[75] Application of IL-6 or coadministration of IFNγ and NGF promote neuronal differentiation of cultured PC12 cells.[76,77] CNTF accentuates the differentiation of sensory neuroblasts from cultured neural crest progenitors, while SCF promotes the elaboration of chick sensory neuronal progenitors; this latter developmental action is blocked by selected neurotrophins.[61,66]

SCF enhances the viability of embryonic dorsal root ganglia neuroblasts in vitro, while LIF promotes the differentiation of peripheral sensory neurons from cultured embryonic dorsal root ganglia cells, with enhancement of neuropeptide expression within these differentiated sensory populations.[61,78,79] OM, when added together with CNTF and LIF, induces a change in neurotransmitter expression and in the profile of early-immediate gene responses in cultured ciliary neurons.[80] For rat sympathetic neuroblasts in vitro,

CNTF or LIF potentiates apoptosis in the embryonic and neonatal period, but this developmental effect switches to positive trophic actions during later postnatal life.[81] After NGF withdrawal, IFN α/β prevents apoptosis of cultured sympathetic neurons.[82] The neuropoietic cytokines, including CNTF, LIF, OM, and CT-1, upregulate cholinergic expression in cultured rat sympathetic neurons, and also modulate neuropeptide patterns and early-immediate gene responses.[54,80] IL-6 and IL-11 regulate a subset of these maturational responses, including the enhancement of substance P (SP) expression (IL-11), and the elaboration of voltage-dependent sodium channels in PC12 cells (IL-6).[61,83,84] IFNγ downregulates the IL-1-induced expression of SP in cultured sympathetic ganglion neurons.[85] In addition, TGFβ1 and 2 promote neuronal survival and the expression of SP within cultured neonatal rat dorsal root ganglia.[8] Further, erythropoietin increases monoamine levels in PC12 cells, and IL-2 enhances neurite outgrowth of cultured sympathetic neurons.[57,86]

Regulation of Cellular Viability During Development

During the evolution of the vertebrate head, neural crest cells emigrate from the dorso- caudal neural tube (hindbrain) and ensheathe the primordial mesodermal mesenchyme to form brachial arches, the template for later construction of mature craniofacial structures.[9] Neural crest cells are produced in discontinuous fashion and carry positional information that allows the orderly formation of consecutive rhombomeres (eight) between the mid- and hindbrain boundaries. During chick embryogenesis, BMP4 is selectively expressed in the dorsal regions of odd-numbered rhombomeres (r3 and 5) and induces segmental apoptosis.[9,87] Through collaborative interactions with adjacent (even-numbered) rhombomeres, BMP4 exhibits significant autoregulation and induces the expression of the transcriptional activator, Msx2.[9,88] Through a BMP4-independent mechanism within r3, Krox-20 is upregulated and there is inhibition of expression of the BMP-binding factor, follistatin.[9] Though cooperative interactions with adjacent rhombomeres is necessary to induce segmental apoptosis, cell culture studies have demonstrated that only r3 and 5 retain the intrinsic susceptibility to BMP4-mediated apoptotic signals.[9,87,88] Further, BMPs are coexpressed within the developing dorsomedian telencephalon at the time of cessation of cell proliferation and an increase in programmed cell death.[89] These cellular events are associated with an upregulation of Msx1 and the concurrent downregulation of Bf1, a forkhead/winged helix transcription factor essential for continued telencephalic proliferation.[89] Therefore, BMPs sanction regional apoptosis during development by activation of specific transcription factors and through collaborative interactions with additional cellular signals.

BMPs also exert selective effects on the survival and terminal differentiation of PNS and brain-derived neuronal progenitor populations during later developmental stages. These experimental observations are consistent with studies of BMP ligand and receptor transcript expression that demonstrate persistent, regional patterns of expression throughout neural development.[7,36,90] BMP7 enhances the expression of specific neural cell adhesion molecules. In concert with activin A, BMP4 potentiates levels of *VGF*, a neural-specific transcript and also augments terminal neuronal differentiation of PC12 cells.[91] Prolonged exposure of EGF-generated embryonic multipotent SVZ progenitor cells to the BMPs results in the later enhancement of cell number and terminal neuritic outgrowth of neuronal progenitor species.[7,35] Further, solitary application of BMPs from four factor subgroups and specific heterodimers (BMP2/6) to more lineage-restricted primary dissociated cells derived from several late-embryonic murine regional neuronal progenitor populations promotes neuronal survival and neurite outgrowth.[7,35] For example, application of BMP2 to lineage-restricted embryonic ventral mesencephalic neuronal subpopulations in vitro potentiates cellular survival and neurite outgrowth of tyrosine hydroxylase- immunoreactive neurons.[7,35,92] Therefore, several different TGFβ related factors (TGFβ2 and 3, glial cell-derived neurotrophic factor, and BMP2) may differentially regulate the elaboration of discrete ventral mesencephalic dopaminergic (DA) neuronal subpopulations through actions on multiple cellular processes.[93,94] Thus, cumulative experimental studies suggest that within the developing brain, BMPs exhibit multiple cellular roles and act upon diverse neural progenitor species. Further, unlike other cytokine families, BMPs may exert these developmental actions by promoting the exit of progenitor species from cell cycle, with subsequent cellular differentiation or apoptosis.

The tumor necrosis factor (TNF) subclass of the hemopoietins includes several associated ligands and receptor subunits, such as the low-affinity neurotrophin receptor (p75^{LNTR}) and APO-1 (Fas)/ CD 95.[95-99] This integrated factor family utilizes sphingomyelinase/ ceramide-activated protein kinase pathways and multiple protease cascades that are positively and negatively modulated by accessory proteins that are themselves regulated by multiple environmental cues. TNFα is synthesized by specific neuronal subpopulations in murine brain, and neuronal receptors utilize several associated cytokines that mediate inflammatory reactions, including IL-1, -2, -6 and IFNγ to influence numerous neuronal and neuroendocrine response parameters.[7,100] Within T cell subpopulations, Fas mediates apoptosis following cell activation and subsequent engagement of the Fas-ligand (autocrine suicide).[100-105] Although Fas-ligand is not normally expressed at significant levels in the murine central nervous system, ligand transcripts are transiently elevated following transient global ischemia.[106]

Glial Cell-Derived Neurotrophic Factor Family

Glial cell-derived neurotrophic factor (GDNF) and the closely related family member, neurturin exhibit a distinct structural similarity to members of the TGFβ superfamily, but also share a basic cysteine knot structure with nerve growth factor and platelet-derived neurotrophic factor.[107-110] However, the receptor signaling motifs utilized by this novel cytokine family more closely resemble those utilized by specific gp130 heterodimeric receptor-associated hemopoietins.[107] Thus, both factors signal through distinct soluble or membrane-associated TGFβ-related neurotrophic factor receptors (TrnRs, a misnomer: TrnR1, GDNFRα; TrnR2, neurturin receptor, NTNRα) linked to the plasma membrane by glycosyl phosphatidylinositol residues, analogous to the arrangement with the CNTFα receptor, and by an additional common receptor subunit, Ret, the product of the *c-ret* protooncogene, that undergoes tyrosine phosphorylation following direct ligand binding.[107,111-113] During hematopoiesis, RET mRNA is present in early CD34$^+$ hematopoietic progenitors and is upregulated during myelomonocytic maturation. GDNFRα is also expressed in bone marrow stromal cells, fibroblasts and osteoblasts, all known to support normal hematolymphoid development.[114] Further, GDNF has been shown to bind transformed myelogenous blast cells and to retard clonogenic growth and potentiate monocytic maturation.[114]

GDNF and neurturin exhibit partially overlapping and distinct biological spectrums of action in the developing nervous system.[110-113] During neural development, GDNF mRNA is expressed within late embryonic and early postnatal striatal and ventral limbic dopaminergic target areas, the cerebellar anlage and developing Purkinje cells, the olfactory tubercle, trigeminal brainstem nuclei, the thalamus and within the dorsal horn and Clarke's column of the spinal cord.[107-109]

Within the embryonic peripheral nervous system, transcripts for GDNF are present within specific myogenic subpopulations, Schwann cells and peripheral neurons.[107-109] At birth, GDNF homozygous null mutants exhibit significant deficits in the elaboration of cranial sensory, sympathetic and enteric neurons, with minor losses of spinal motor neurons and no significant changes within dopaminergic neuronal subpopulations; *c-ret* knockout mice exhibit similar developmental profiles.[107] GDNF maintains the survival of motor neurons in culture, with trophic effects that exceed those of neurotrophin-3, prevents cell death after axotomy-induced injury and represents a target-derived growth factor that is present in both muscle and peripheral nerves.[107-109] In addition, GDNF has significant trophic actions on sympathetic, ciliary, spinal sensory and nodose neurons. Further, sensitivity to the biological effects of GDNF decreases with age in sympathetic, parasympathetic, and proprioceptive sensory neurons, but increases with age in exteroceptive and cutaneous sensory neurons.[109] Both GDNF and neurturin support the survival of sympathetic, sensory, nodose and dorsal root ganglia neurons.[107-113] GDNF also promotes the development of adrenergic neurons from murine neural crest cultures with potentiation of expression of tyrosine hydroxylase and selective expression of intermediate- stage neuronal lineage markers, indicating the presence of a critical period for trophic actions.[115] Members of the GDNF factor family thus have developmental actions on cultured murine neural crest cells that are distinct from those of the BMPs and other TGFβ superfamily cytokines.

Expression of Cytokine Signaling Molecules During Neural Development

Transcripts for IL-3 are preferentially expressed during early embryonic development (E13, E16) and this may reflect the essential role of this hemopoietin in CNS progenitor cell survival and proliferation (Table 1.3).[7] IL-12 mRNA, by contrast, exhibits highest expression during late postnatal (P18) development and in the adult; this developmental profile is consonant with the actions of this cytokine on terminal axodendritic process outgrowth within multiple brain regions.[7] In addition, the expression of transcripts for IL-5 and -7 are maximal in late embryonic (E16) and early postnatal (PN1) brain, consistent with the role of these interleukins in more intermediate stages of neurogenesis.[7,42] Examination of transcript patterns for SCF and its cognate receptor, Kit, have been studied in

*Table 1.3. Selected expression of hematolymphopoietic cytokine
ligand and receptor transcripts during CNS development*

LIGANDS:	E13	E16	PN1	PN18	AD
IL-3	+++	++	+	+	+/-
IL-4	+	++	+	-	-
IL-5	-	+/-	+	+++	-
IL-7	+	+++	+++	+	+
IL-12	-	-	+	++	+++
Epo	-	-	++	++	+
CSF1	n.d.	++	++	++	++
SCF	+	++	++	++	+
Receptors:					
IL-4Rα	+	++	+	-	-
IL-5Rα	-	+	++	++	-
IL-7Rα	+	++	++	+	+
c-Fms	+	++	++	+++	++
c-Kit	+	++	++	+	+

greater detail and reveal intricate complementary regional and cellular profiles that highlight the distinctive roles of SCF in both early stages of neural crest differentiation and in more advanced stages of regional CNS neuronal terminal differentiation.[45,66,116-118] There is a close correspondence in both temporal and spatial expression of transcripts for other hemopoietin ligand-receptor pairings. Following IL-7 application, single cell analysis of primary embryonic murine and rodent neuroblast populations has revealed direct cytokine cellular actions on nascent neural populations.[47] Additional studies have shown that IL-7 is preferentially expressed in glial subpopulations, while the IL-7 receptor subunit is preferentially found on neuronal subpopulations.[47] In vivo, IL-7 may thus be secreted by specific regional glial cells and may target contiguous nascent neuroblast subpopulations. Other experimental investigations have found that application of IL-7 to primary cultured embryonic cortical neuroblast populations causes selective and rapid expression of *c-myc* transcripts and phosphorylation of the Src family protein, p59[fyn].[47] Comparative analysis of homozygous null mutations for hemopoietin ligands and receptors may suggest the existence of additional, yet undiscovered ligands that mediate developmental signaling within specific temporospatial settings. As an example, there is a profound discrepancy between the regional and cellular deficits of CNTFRα-/ -(severe motor neuron defects and periventricular dilatation) and LIFRβ-/-(profound loss of astroglial lineage species) mice and those of the homozygous null mutations for the corresponding ligands, CNTF and LIF, which display substantially fewer abnormalities.[54,119,120] These observations strongly suggest the presence of additional ligands that are active during embryonic development within distinct spatiotemporal contexts.

TGFβ ligand and receptor transcripts and proteins are expressed throughout neural development, consistent with the complex spectrum of the effects of these cytokines on multiple neural progenitor species and on more differentiated cellular progeny.[121,122] TGFβ1 mRNAs are present in early neural crest cells and in cultured brain macrophages.[7] In addition, TGFβ1 protein is expressed in the meninges, and TGFβ isoforms are differentially expressed during embryogenesis.[7,121,122] Further, TGFβ2 and 3 transcripts are selectively upregulated during early embryonic life and are also present during later developmental stages in the brain and spinal cord.[123] Although TGFβ2 shows preferential expression in the cerebral cortex, TGFβ3 is predominantly expressed in the olfactory bulb.[54] Oligodendrocytes and Schwann cells both produce TGFβ1 and proliferate in response to its presence.[7] All three TGFβ isoforms are expressed in cultured mouse astrocytes, while TGFβ2 is secreted in latent form.[7]

Transcripts for BMP ligands (BMPs 2-7) are expressed throughout brain development with a complex temporospatial pattern of expression (Table 1.4).[7,35,36,90] At a cellular level, transcripts for BMP2 and BMP7 are preferentially localized to cultured neurons, astrocytes, and microglia, while those for BMP4 are localized to bipotent oligodendroglial-astroglial progenitor cells and to oligodendrocytes.[7,38,124] BMP6 transcripts are expressed in rat radial glia and neurons during the perinatal and early postnatal periods.[125,126] In addition, transcripts for BMP4 are present in early diencephalic neu-

Table 1.4A. BMP transcript expression during CNS development

	E13	E16	R19	PN4	PN14
BMP2	+	+	+	++	++
BMP3	+	++	+++	+++	+++
BMP4	+++	++	++	+++	++++
BMP5	+	+	+	++	++++
BMP6	++	+++	++++	++++	++
BMP7	++++	+++	+++	+++	++

Table 1.4B. Regional BMP transcript expression during CNS development

BMP6	E16	E18	PN1	PN4	AD
Brainstem	+++	+++	++	++	+++
Cerebellum	++	++	+++	+++	++
Cortex	+	++	++	++	+
Hippocampus	++	+++	+++	++	+
Striatum	+	+	++	++	++

roepithelium in association with Rathke's pouch, and additional transcripts for BMP7 and 8a are expressed within a hippocampal cDNA library.[5,7] Transcripts for BMP2, 4 and 6 are expressed in the dorsal neural tube and those for BMP4 and BMP7 in the epidermal ectoderm during progressive stages of neural development.[8] Finally, transcripts for growth/differentiation factor (GDF) 5 and 10 have been found within the rodent brain.[90] Transcripts for BMP type I receptors, BMPRIA and IB, are present at significant levels in murine periventricular generative zones during early embryogenesis, with a partially overlapping developmental pattern throughout embryonic life.[90,127] BMPRIA, IB and BMPRII mRNAs are also expressed during early and intermediate stages of cerebral cortical development.[90] Expression of these BMP type I and II receptor subunit transcripts are highest during embryonic development and decline in postnatal life.[90] In addition, BMP receptor subunit proteins are localized to ventricular and subventricular zones as early as embryonic day E12. Embryonic subventricular zone multipotent and oligopotent progenitor cells express a full complement of transcripts for BMP type I and II receptors and for selected BMP ligands (BMP4 and BMP7 but not BMP2), suggesting a spectrum of early autocrine and paracrine signaling events.[7,35,36] BMP developmental analysis is further complicated by the fact that although BMP ligands and receptors display partially overlapping expression patterns, homozygous null mutants often display distinct developmental phenotypes.[5,23,128] Transcripts for ActRI and ActRII are also present during early neural development.[90]

Conclusions

Cytokines from three distinct factor families not previously thought to be involved in neural development, the hemopoietin and TGFβ superfamilies and the GDNF family, mediate a diverse range of cellular events during neurulation, morphogenesis and early, intermediate and later stages of neural lineage elaboration and terminal differentiation. These cytokines exhibit complementary, cooperative and antagonistic signaling interactions during progressive developmental stages. Analysis of the patterns of expression of cytokine ligands, receptor subunits, intracellular transduction molecules, extracellular binding proteins and targeted homozygous null mutations have demonstrated that these growth factors exert their developmental actions through autocrine, paracrine and cooperative signaling loops and through the establishment of morphogenetic gradients. Further, these factors act on a range of multipotent, oligopotent and more lineage-restricted progenitor species to influence cellular proliferation, survival, regional apoptosis, lineage commitment, phenotypic maturation and to potentiate the expression of neurotransmitters, neuropeptides, ion and ligand-gated channels, synaptic terminal proteins and cell adhesion molecules. These experimental studies suggest that the cellular mechanisms utilized by these cytokines during sequential stages of hematolymphopoiesis will have significant parallels to those active during central and peripheral nervous system development. Null mutations of cytokine signaling components frequently lead to minimal phenotypic changes in the brain, since there is substantial redundancy and pleiotropy in cytokine actions in the central nervous system. The advent of targeted knockouts of these cytokine signaling components within specific developmental periods, anatomical regions and progenitor species will allow more detailed definitions of the integrated cellular and molecular mechanisms by which these growth factors exert their biological actions. Further, identification of the detailed cellular, developmental and clinical consequences of perturbations of these cytokine signaling pathways will provide insights into the pathogenesis of neuroimmunodegeneration in mice and men.

References

1. Mori H, Tanaka R, Yoshida K et al. Immunological analysis of the rats with anterior hypothalamic lesions. J Neuroimmunol 1993; 48:45-52.
2. Gushchin GV, Jakavleva EE, Kataeva GV et al. Muscarinic cholinergic receptors of rat lymphocytes: Effect of antigen stimulation and local brain lesion. Neuroimmunomod 1994; 1:259-264.
3. Heninger GR. Neuroimmunology of stress. In: Friedman MJ, Charnry DS, Deutch AY, eds. Neurobiological and Clinical Consequences of Stress: From Normal Adaptation to PTSD. Philadelphia: Lippincott-Raven Publishers, 1995:381-401.
4. Felten DL, Cohen N, Ader R et al. Central circuits involved in neural-immune interactions. In: Ader R, Felten DL, Cohen N, eds. Psychoneuroimmunology, 2nd ed. New York: Academic Press Inc., 1991:3-25.
5. Hogan BLM. Bone morphogenetic proteins: Multifunctional regulators of vertebrate development. Genes Develop 1996; 10:1580-1984.
6. Kingsley DM. The TGF-beta superfamily: New members, new receptors, and new genetic tests of function in different organisms. Genes Develop 1994; 8:133-146.
7. Mehler MF, Kessler JA. Cytokines and neuronal differentiation. Crit Rev Neurobiol 1995; 9:419-446.
8. Tanabe Y, Jessell TM. Diversity and pattern in the developing spinal cord. Science 1996; 274:1115-1123.
9. Graham A, Koentges G, Lumsden A. Neural crest apoptosis and the establishment of craniofacial pattern: An honorable death. Mol Cell Neurosci 1996; 8:76-83.
10. Massague J. TGF beta signaling: Receptors, transducers, and Mad proteins. Cell 1996; 85:947-950.
11. Wozney JM, Rosen V, Celeste AJ et al. Novel regulators of bone formation. Science 1988; 242:1528-1534.
12. Lecuit T, Brook WJ, Ng M et al. Two distinct mechanisms for long-range patterning by *Decapentaplegic* in the Drosophila wing. Nature 1996; 381:387-393.

13. Hemmati-Brivanlou A, Melton D. Vertebrate embryonic cells will become nerve cells unless told otherwise. Cell 1997; 88:13-17.

14. Wilson PA, Hemmati-Brivanlou A. Induction of epidermis and inhibition of neural fate by BMP4. Nature 1995; 376:331-333.

15. Hemmati-Brivanlou A, Melton DA. Inhibition of activin receptor signaling promotes neuralization of *Xenopus*. Cell 1994; 77:273-281.

16. Lemaire P, Kodjabachian L. The vertebrate organizer: Structure and molecules. Trends Genet 1996; 12:525-531.

17. Piccolo S, Sasai Y, Lu B et al. Dorsoventral patterning in *Xenopus*: inhibition of ventral signals by direct binding of chordin to BMP4. Cell 1996; 86:589-598.

18. Zimmerman LB, DeJesus-Escobar JM, Harland RM. The Spemann organizer signal noggin binds and inactivates bone morphogenetic protein 4. Cell 1996; 86:599-606.

19. Holley SA, Neul JL, Attisano L et al. The *Xenopus* dorsalizing factor noggin ventralizes *Drosophila* embryos by preventing Dpp from activating its receptor. Cell 1996; 86:607-617.

20. Hemmati-Brivanlou A, Kelley OG, Melton DA. Follistatin, an antagonist of activin, is expressed in the Spemann organizer and displays direct neuralizing activity. Cell 1994; 77:283-295.

21. Lumsden A, Krumlau FR. Patterning the vertebrate neuraxis. Science 1996; 274:1109- 1115.

22. Liem KF, Tremml G, Roelink H et al. Dorsal differentiation of neural plate cells induced by BMP-mediated signals from epidermal ecoderm. Cell 1995; 82:969-979.

23. Winnier G, Blessing M, Labosky PA et al. Bone morphogenetic protein 4 is required for mesoderm formation and patterning in the mouse. Genes Develop 1995; 9:2105-2116.

24. Roelink H. Tripartite signaling of pattern: Interactions between Hedgehogs, BMPs and Wnts in the control of vertebrate development. Curr Opin Neurobiol 1996; 6:33-40.

25. Biehs B, Francois V, Bier E. The *Drosophila short gastrulation* gene prevents Dpp from autoactivating and suppressing neurogenesis in the neuroectoderm. Genes Develop 1996; 10:2922-2934.

26. Mehler MF, Kessler JA. Stem cells and neural development. In: Adelman G, Smith B, eds. Encyclopedia of Neuroscience. Amsterdam: Elsevier Science (in press).

27. Weiss S, Reynolds BA, Vescovi A et al. Is there a neural stem cell in the mammalian forebrain? Trends Neurosci 1996; 19:387-393.

28. Richards LJ, Kilpatrick TJ, Dutton R et al. Leukemia inhibitory factor or related factors promotes the differentiation of neuronal and astrocytic precursors within the developing murine spinal cord. Eur J Neurosci 1996; 5:291-300.

29. Pfeiffer SE, Warrington AE, Bansal R. Oligodendrocyte and its many cellular processes. Trends Cell Biol 1993; 3:191-198.

30. Barres BA, Schmid R, Sendtner M et al. Multiple extracellular signals are required for long-term oligodendrocyte survival. Development 1993; 8:283-295.

31. Louis J-C, Muir D, Varon S. Autocrine inhibition of mitotic activity in cultured oligodendrocyte-type 2 astrocyte (O-2A) precursor cells. GLIA 1992; 6:30-38.

32. McKinnon RD, Piras G, Ida JA et al. A role for TGF-beta in oligodendrocyte differentiation. J Cell Biol 1993; 121:1397-1407.

33. Van Meir EG. Cytokines and tumors of the central nervous system. GLIA 1995; 15:264-288.

34. Marmur R, Mehler MF, Mabie PC et al. Characterization of pre-O2A progenitor cultures from mouse embryonic subventricular zone progenitor cells. Soc Neurosci Abstr 1995; 21:287.

35. Mehler MF, Marmur R, Gross RE et al. Cytokines regulate the cellular phenotype of developing neural lineage species. Int J Develop Neurosci 1995; 13:213-240.

36. Gross RE, Mehler MF, Mabie PC et al. Bone morphogenetic proteins promote astroglial lineage commitment by mammalian subventricular zone progenitor cells. Neuron 1996; 17:595-606.

37. Franklin RJM, Blakemore WF. Glial-cell transplantation and plasticity in the O-2A lineage- implications for CNS repair. Trends Neurosci 1995; 18:151-156.

38. Mabie PC, Mehler MF, Marmur R et al. Oligodendroglial and astroglial differentiation during development and remyelination. Ann Neurol 1996; 40:546.

39. Giesen K, Hummel T, Stollework A et al. Glial development in the *Drosophila* CNS requires concomitant activation of glial and repression of neuronal differentiation genes. Development 1997; 124: 23078-2311.

40. Edwards MA, Yamamoto M, Caviness VS. Organization of radial glia and related cells in the developing murine CNS. An analysis based upon a new monoclonal antibody marker. Neuroscience 1990; 36:121-144.

41. Mehler MF, Mabie PC, Marmur R et al. Differential regulation of radial glia and astroglial lineage elaboration from embryonic subventricular zone progenitor cells by leukemia inhibitory factor receptor activation and bone morphogenetic proteins. Soc Neurosci Abst 1996; 22:285.

42. Mehler MF, Rozental R, Dougherty M et al. Cytokine regulation of neuronal differentiation of hippocampal progenitor cells. Nature 1993; 362:62-65.

43. Keegan AD, Nelms K, Wang L-M et al. Interleukin 4 receptor: Signaling mechanisms. Immunology Today 1994; 14:423-432.

44. Mabie PC, Mehler MF, Marmur R et al. The developmental neurotrophic potential of an interleukin subset that regulates hematolymphoid development and signals through a common gamma receptor subunit. Neurology 1995; 45:A335.

45. Motro B, Van der Kooy D, Rossant J et al. Contiguous patterns of *c-kit* and *steel* expression: Analysis of mutations at the *W* and *Sl* loci. Development 1991; 113:1207-1221.

46. Michaelson MD, Bieri P, Mehler MF et al. CSF1 deficiency in mice results in abnormal brain development. Development 1996; 122: 2661-2672.

47. Michaelson MD, Mehler MF, Xu H et al. Interleukin 7 is trophic for embryonic neurons and is expressed in developing brain. Develop Biol 1996; 179:251-263.

48. Du X, Stull ND, Iacovitti L. Brain-derived neurotrophic factor works coordinately with partner molecules to initiate tyrosine hydroxylase expression in striatal neurons. Brain Res 1995; 680:229-233.

49. Kato AC, Lindsay RM. Overlapping and additive effects of neurotrophins and CNTF on cultured human spinal cord neurons. Exp Neurol 1994; 130:196-210.

50. Michikawa M, Kikuchi S, Kim SU. Leukemia inhibitory factor-mediated increase of choline acetyltransferase activity in mouse spinal cord neurons in culture. Neurosci Lett 1992; 140:75-77.

51. Kamegai M, Niijima K, Kunishita T et al. Interleukin 3 as a trophic factor for central cholinergic neurons *in vitro* and *in vivo*. Neuron 1990; 2:429-436.

52. Konishi Y, Chui D, Hirose H et al. Trophic effects of erythropoietin and other hematopoietic factors on central cholinergic neurons *in vitro* and *in vivo*. Brain Res 1993; 609:29-35.

53. Tabira T, Konishi Y, Gallyas F Jr. Neurotrophic effect of hematopoietic cytokines on cholinergic and other neurons *in vitro*. Int J Dev Neurosci 1995; 13:241-252.

54. Henderson CE. Role of neurotrophin factors in neural development. Curr Opin Neurobiol 1996; 6:64-70.

55. Snyder SH, Sabatini DM. Immunophilins in the nervous system. Nature Med 1995; 1:32- 37.

56. Gottschalk AR, Boise LH, Thompson CB et al. Identification of immunosuppressant- induced apoptosis in murine B-cell line and its prevention by *bcl-x* but not *bcl-2*. Proc Natl Acad Sci USA 1994; 91:7350-7354.

57. Plata-Salaman CR. Immunoregulators in the nervous system. Neurosci Biobehav Rev 1991; 15:185-215.

58. Pliophys AV. Expression of the 210 kDa neurofilament subunit in cultures of nervous system from normal and trisomy 16 mice: Regulation by interferon. J Neurosci 1988; 85:209-222.

59. Barish M, Mansdorf NB, Raissdana SS. Gamma-interferon promotes differentiation of cultures cortical and hippocampal neurons. Develop Biol 1991; 144:412-423.

60. Erkman L, Wuarin L, Cadelli D et al. Interferon induces astrocyte maturation causing an increase in cholinergic properties of cultured spinal cord cells. Develop Biol 1989; 132:375-388.

61. Patterson PH. The emerging neuropoietic cytokine family: First CDF/LIF, CNTF and IL-6; next ONC, MGF, GCSF? Curr Opin Neurobiol 1992; 2:94-97.

62. Campfield LA, Smith FJ, Guisez Y et al. Recombinant mouse OB protein: Evidence for a peripheral signal linking adiposity and central neural networks. Science 1995; 269:546-549.

63. Gainsford T, Wilson T, Metcalf D et al. Leptin can induce proliferation, differentiation and functional activation of hematopoietic cells. Proc Natl Acad Sci USA 1996; 93:14564-14568.

64. Bennett BD, Solar GP, Yuan JQ et al. A role for leptin and its cognate receptor in hematopoiesis. Curr Biol 1996; 6:1170-1180.

65. Cioffi JA, Shafer AW, Zupancic TJ et al. Novel B219/OB receptor isoforms: Possible role of leptin in hematopoiesis and reproduction. Nature Med 1996; 2:585-589.

66. Langtimim-Sedlak CJ, Schroeder B, Saskowski JL et al. Multiple actions of stem cell factor in neural crest differentiation *in vitro*. Develop Biol 1996; 174:345-359.

67. Groves A, Anderson DJ. Role of environmental signals and transcriptional regulators in neural crest development. Develop Genet 1996; 18:64-72.

68. Shah NM, Groves AK, Anderson DJ. Alternative neural crest cell fates are instructively promoted by TGF beta superfamily members. Cell 1996; 85:331-343.

69. Shah NM, Marchionni MA, Isaacs I et al. Glial growth factor restricts mammalian neural crest cells to a glial fate. Cell 1994; 77:349-360.

70. Reissmann E, Ernsberger U, Francis-West PH et al. Involvement of bone morphogenetic protein 4 and bone morphogenetic protein 7 in the differentiation of the adrenergic phenotype in developing sympathetic neurons. Development 1996; 122:2079-2088.

71. Varley JE, Wehby RG, Rueger DL et al. Number of adrenergic and islet-1 immunoreactive cells is increased in avian trunk neural crest cultures in the presence of human recombinant osteogenic protein-1. Develop Dynamics 1995; 203:434-447.

72. Varley JE, Maxwell GD. BMP-2 and BMP-4, but not BMP-6, increase the number of adrenergic cells which develop in quail trunk neural crest cultures. Exp Neurol 1996; 140:84-94.

73. Fann M-J, Patterson PH. Depolarization differentially regulates the effects of bone morphogenetic protein (BMP)-2, BMP-6 and activin A on sympathetic neuronal phenotype. J Neurochem 1994; 63:2074-2079.

74. Lein P, Johnson M, Guo X et al. Osteogenic protein-1 induces dendritic growth in rat sympathetic neurons. Neuron 1995; 15:597-605.

75. Ip NY, Boulton TG, Li Y et al. CNTF, FGF and NGF collaborate to drive the terminal differentiation of MAH cells into postmitotic neurons. Neuron 1994; 13:443-455.

76. Improta T, Salvatore AM, DiLuzio A et al. IFN gamma facilitates NGF-induced neuronal differentiation in PC12 cells. Exper Cell Res 1988; 179:1-9.

77. Wu YY, Bradshaw RA. Induction of neurite outgrowth by interleukin-6 is accompanied by activation of STAT3 signaling pathway in a variant PC12 cell (E2) line. J Biol Chem 1996; 271:13023-13032.

78. Murphy M, Reid K, Hilton DJ et al. Generation of sensory neurons is stimulated by leukemia inhibitory factor. Proc Natl Acad Sci USA 1991; 99:3498-3501.

79. Bamber BA, Masters BA, Hoyle GW et al. Leukemia inhibitory factor induces neurotransmitter switching in transgenic mice. Proc Natl Acad Sci USA 1994; 91:7839-7843.

80. Rao MS, Symes A, Malik N et al. Oncostatin M regulates VIP expression in a human neuroblastoma cell line. NeuroReport 1992; 3:865-868.

81. Kessler JA, Ludlam WH, Freidin MM et al. Cytokine-induced programmed death of cultures sympathetic neurons. Neuron 1993; 11:1123-1132.

82. Chang JY, Martin DP, Johnson FM et al. Interferon suppresses sympathetic neuronal cell death caused by nerve growth factor deprivation. J Neurochem 1990; 55:436-445.

83. Patterson PH. Neuronal growth and differentiation factors and synaptic plasticity. In: Bloom FE, Kupfer DJ, eds. Psychopharmacology: The Fourth Generation of Progress, New York: Raven Press, 619-629.

84. Fann M-J, Patterson PH. Neuropoietic cytokines and activin A differentially regulate the phenotype of cultured sympathetic neurons. Proc Natl Acad Sci USA 1994; 91:43-47.

85. Olsson T. Cytokines in neuroinflammatory disease: Role of myelin autoreactive T cell production of interferon-gamma. J Neuroimmunol 1992; 40:211-218.

86. Masuda S et al. Functional erythropoietin receptor of the cells with neural characteristics. J Biol Chem 1993; 268:11208-11216.

87. Graham A, Francis-West P, Brickell P et al. The signalling molecule BMP4 mediates apoptosis in the rhombencephalic neural crest. Nature 1994; 387:684-686.

88. Marazzi G, Wang, Y, Sassoon D. Msx2 is a transcriptional regulator in the BMP4-mediated programmed cell death pathway. Develop Biol 1997; 186:127-138.

89. Furuta Y, Piston DW, Hogan BL. Bone morphogenetic proteins (BMPs) as regulators of dorsal forebrain development. Development 1997; 124:2203-2212.

90. Soderstrom S, Bengtsson H, Ebendahl T. Expression of serine/threonine kinase receptors including the bone morphogenetic factor type II receptor in the developing and adult rat brain. Cell Tissue Res 1996; 286:269-279.

91. Paralkar VM, Weeks BS, Yu Y et al. Recombinant human bone morphogenetic protein 2B stimulates PC12 cell differentiation: Potentiation and binding to type IV collagen. J Cell Biol 1992; 119:1721-1728.

92. Mabie PC, Mehler MF, Kessler JA. Effects of bone morphogenetic protein 2 on rat embryonic ventral mesencephalon *in vitro*. Soc Neurosci Abstr 1994; 20:665.

93. Beck KD, Valverde J, Alexi T et al. Mesencephalic dopaminergic neurons protected by GDNF from axotomy-induced degeneration in the adult brain. Nature 1995; 373:339-341.

94. Poulsen KT, Armanini MP, Klein RD et al. TGF beta 2 and TGF beta 3 are potent survival factors for midbrain dopaminergic neurons. Neuron 1994; 13:1245-1252.

95. Kolesnick R, Golde DW. The sphingomyelin pathway in tumor necrosis factor and interleukin-1 signalling. Cell 1994; 77:325-328.

96. Armitage RJ. Tumor necrosis factor receptor superfamily members and their ligands. Curr Opin Immunol 1994; 6:407-413.

97. Squier MKT, Cohen JJ. Cell-mediated cytotoxic mechanisms. Curr Opin Immunol 1994; 6:447-452.

98. Smith CA, Farrah T, Goodwin RG. The TNF receptor superfamily of cellular and viral proteins: activation, costimulation and death. Cell 1994; 76:959-962.

99. Heller RA, Kronke M. Tumor necrosis factor receptor-mediates signaling pathways. J Cell Biol 1994; 126:5-9.

100. Bartfai T, Schultzberg M. Cytokines in neuronal cell types. Neurochem Int 1993; 22:435- 444.

101. Ju S-T, Panka DJ, Cui H et al. Fas (CD95)/FasL interactions required for programmed cell death after T-cell activation. Nature 1995; 373:444-448.

102. Brunner T, Mogil RJ, LaFace D et al. Cell-autonomous Fas (CD95)/Fas ligand interactions mediates activation-induced apoptosis in T-cell hybridomas. Nature 1995; 373:441-444.

103. Dhein J, Walczak H, Baumler C et al. Autocrine T-cell suicide mediated by APO- 1/(Fas/CD95). Nature 1995; 373:438-441.

104. Strasser A. Death of a T cell. Nature 1995; 373:385-386.

105. Kagi D, Vignaux F, Ledermann B et al. Fas and perforin pathways as major mechanisms of T cell-mediated cytotoxicity. Science 1994; 265:528-530.

106. Matsuyama T, Hata R, Tagaya M et al. Fas antigen mRNA induction in postischemic murine brain. Brain Res 1994; 657:342-346.

107. Lindsay RM, Yancopoulos GD. GDNF in a bind with known orphan: Accessory implicated in new twist. Neuron 1996; 17:571-574.

108. Unsicker, K. GDNF: A cytokine at the interface of TGF-β and neurotrophins. Cell Tissue Res 1996; 286:175-178.

109. Nosrat CA, Tomac A, Linquist E et al. Cellular expression of GDNF mRNA suggests multiple functions inside and outside the nervous system. Cell Tissue Res 1996; 286:191-207.

110. Kotzbauer PT, Lampe PA, Heukeroth RO et al. Neurturin, a relative of glial-cell-line- derived neurotrophic factor. Nature 1996; 384: 467-470.

111. Baloh RH, Tansey MG, Golden JP et al. TrnR2, a novel receptor that mediates neurturin and GDNF signaling through Ret. Neuron 1997; 18:793-802.

112. Klein RD, Sherman D, Ho W-H. A GPI-linked protein that interacts with Ret to form a candidate neurturin receptor. Nature 1997; 387:717-721.

113. Buj-Bello A, Adu J, Pinon LGP et al. Neurturin responsiveness requires a GPI-linked receptor and the Ret receptor tyrosine kinase. Nature 1997; 387:721-724.

114. Gattei V, Celetti A, Cerrato A et al. Expression of the Ret receptor tyrosine kinase and GDNFR-alpha in normal and leukemic human

hematopoietic cells and stromal cells of the bone marrow microenvironment. Blood 1997; 89:2925-2937.

115. Maxwell GD, Reid K, Elefanty A et al. Glial cell line-derived neurotrophic factor promotes the development of adrenergic neurons in mouse neural crest cultures. Proc Natl Acad Sci USA 1996; 93:13274-13279.

116. Keshet E, Lyman SD, Williams DE et al. Embryonic RNA expression patterns of the *c-kit* receptor and its cognate ligand suggest multiple functional roles in mouse development. EMBO J 1991; 10:2425-2435.

117. Morii E, Hirota S, Kim H-M et al. Spatial expression of genes encoding *c-kit* receptors and their ligands in mouse cerebellum as revealed by in situ hybridization. Dev Brain Res 1992; 65:123-129.

118. Orr-Urteger A, Aviv A, Zimmer Y et al. Developmental expression of *c-kit*, a protooncogene encoded by the *W locus*. Development 1990; 109:911-923.

119. Stahl N, Yancopoulos GD. The alphas, betas and kinases of cytokine receptor complexes. Cell 1993; 7:587-590.

120. DeChiara TM, Vejada R, Poveymirou WT et al. Mice lacking the CNTF receptor, unlike mice lacking CNTF, exhibit profound motor neuron deficits at birth. Cell 1995; 83:313-322.

121. Finch CE, Laping NJ, Morgan TE et al. TGF-beta 1 is an organizer of responses to neurodegeneration. J Cell Biochem 1993; 53:314-322.

122. Chalazonitis A, Kalberg J, Twardzik DR et al. Transforming growth factor β has neurotrophic actions on sensory neurons *in vitro* and is synergistic with nerve growth factor. Develop Biol 1992; 152:121-132.

123. Flanders KC, Ludecke G, Engels S et al. Localization and actions of transforming growth factor-beta in the embryonic nervous system. Development 1991; 113:183-191.

124. Mabie PC, Mehler MF, Papavasiliou AK et al. Distinct cellular targets of the bone morphogenetic proteins in neural development. Soc Neurosci Abstra 1995; 21:1542.

125. Schluesener HJ, Meyermann R. Expression of BMP-6, a TGFβ related morphogenetic cytokine, in rat radial glia. GLIA 1994; 12:161-164.

126. Tomizawa K, Matsui H, Kondo E et al. Developmental alterations and neuron-specific expression of bone morphogenetic protein-6 mRNA in rodent brain. Mol Brain Res 1995; 28:122-128.

127. DeWulf N, Verschueren K, Lonnoy O et al. Distinct spatial and temporal expression patterns of two type I receptors for bone morphogenetic proteins during mouse embryogenesis. Endocrinology 1995; 136:2652-2663.

128. Mishini Y, Suzuki A, Ueno N et al. Bmpr encodes a type I bone morphogenetic protein receptor that is essential for gastrulation during mouse embryogenesis. Genes Develop 1995; 9:3027-3037.

CHAPTER 2

Neuroimmunodegeneration Syndromes: Definition and Models

Paul K.Y. Wong and William S. Lynn

Definition

The two major biological systems that must monitor and respond appropriately to environmental changes are the nervous and immune (central and peripheral) systems. These two systems are highly integrated and constantly interact with each other.

Several lines of evidence support the notion that the nervous and immune systems are inextricably linked and functionally connected. First, both the nervous and immune systems are derived from the neural crest.[1] Second, many of the neuropoietic and hemato-lymphopoetic cytokine signals regulate the early development of both these systems (see chapter 1). Third, the central nervous system (CNS) is connected to the peripheral immune system anatomically and physiologically.[2] Fourth, the CNS, which has developed its own local secluded glial immune/defense system, utilizes many of the same proinflammatory cytokines used by the peripheral immune system. Fifth, many neuropeptides or neurotransmitters are produced both by lymphoid cells and CNS cells and can regulate the function of both immune and neural cells. Sixth, both the nervous and immune systems utilize common cytokine signals to communicate with each other and to regulate cell growth and cell death. Seventh, although the cellular and humoral interaction between the central immune and peripheral immune systems are normally restricted in adults by the blood-brain barrier, the ability of the astrocytic support

Neuroimmunodegeneration, edited by Paul K.Y. Wong and William S. Lynn.
© 1998 Springer-Verlag and R.G. Landes Company.

cells in the CNS to quickly differentiate into macrophage like cells, which more readily communicate with the peripheral immune system, greatly amplifies CNS-peripheral immune system interactions.

In postnatal syndromes in which cell injury or impairment of both systems occurs, as in retrovirus infection, the onset of multiple disorders involving both the nervous and immune systems may become rampant. We have previously described such disorders as neuroimmunodegeneration (NID) syndromes.[3] However, because the CNS also has its own unique immune defense system (astroglial and microglial cells) and because neurons are parasitic and very dependent on their local immune/support system for nutrients and redox support and for maintenance of appropriate concentrations of many neurotransmitters,[3-7] impairment or dysfunction of the CNS immune/support cells often leads to early death of the dependent neurons.[5] Thus it seems appropriate to include those neuronal disorders that result from impairment or dysfunction of the CNS immune/support glial system as NID syndromes.

By the above definition, a large number of diseases can be included under the term NID. Table 2.1 presents some examples of NID syndromes. These include the human diseases ataxia-telangiectasia (AT) and human immunodeficiency virus (HIV) infection, in which cell loss occurs in both the immune and nervous systems. Other examples of NID syndromes are Alzheimer's disease (AD), Down syndrome (DS), and Parkinson's disease (PD), in which peripheral immune cell dysfunction or death is minimal but in which neuronal death maybe etiologically associated with dysregulated interactions between neurons and their local activated glial support system. Pathology in the brain of patients with these NID syndromes is usually characterized by the accumulation of cytotoxic or inflammagenic fibrillary and structural peptides produced by the activated astrocytes or the dying neurons.

There are also several examples of animal models of NID (Table 2.1). These include the syndromes observed in ATm knockout (KO) mice and "wasted" mice, in which the fates of both immune and neural cells are markedly altered. Retrovirus animal models of NID include infection by simian immunodeficiency virus (SIV), feline immune deficiency virus (FIV), and *ts1* Moloney murine leukemia (*ts1* MoMuLV), all of which produce disorders that afflict both the peripheral immune and nervous systems, yielding immune and neuronal cell losses. Other murine retrovirus models include mouse AIDS

Table 2.1. Examples of NID syndromes

Diseases	Cell loss or impaired cell functions*	
	CNS	Peripheral immune system
Human		
Ataxia telangiectasia	+++	+++
AIDS and AIDS dementia complex	+++	+++
Alzheimer's disease	+++	+
Down syndrome	+++	+
Parkinson's disease	+++	+
Animal models		
ATm-deficient mouse	+	+++
Wasted mouse	+++	+++
Retrovirus infections		
SIV	++	+++
FIV	+	+++
ts1 MoMuLV	+++	+++
MAIDS	+	+++
CasBrE	+++	±
Transgenic mice overexpressing cytokines in CNS (e.g., TNF, IL-6)	+++	ND

ND, not determined
*Relative extent of dysfunction or loss of neuronal cells or astrocytes in CNS and relative extent of loss or dysfunction of lymphoid cell in the peripheral immune system

(MAIDS), which has detrimental effects on the peripheral immune system but milder effects in the nervous system (chapter 5). The retrovirus isolated from field mice of Lake Casita, California (CasBrE), infects the support cells in the CNS and induces drastic neurological disorders with massive astroglial and microglial reactions, but no degeneration in the peripheral immune system.[8] In transgenic mice, overexpression of proinflammatory cytokines IL-3, IL-6, IFNα/γ, TNFα, or TGFβ specifically in astrocytes results in neural cell death with striking upregulation of astroglial growth and differentiation pathways (see chapter 6). The chronic and massive overexpression of TNFα by astrocytes or neuronal cells in the CNS of transgenic mice also results, at a late stage of the disease, in a massive influx of lymphocytes and macrophages with localized demyelination, axonal destruction, astrogliosis and subsequent neurodegeneration (see chapter 7). This probably represents the combined effect of a high concentration of TNF in the CNS, neurotoxin release by the invading inflammatory cells, and the loss of neuronal support by the highly activated astrocytes.

Thus the models of NID presented above cover a spectrum of syndromes ranging from dysfunction in both nervous and peripheral immune systems to damage to neurons as a result of impairment of the CNS local immune/support cells. Together, such models offer exciting possibilities for understanding the pathogenic mechanisms of human NIDs. Understanding these mechanisms should provide insights that may lead to the development of therapies for these NID syndromes.

Models of NID

Retrovirus-Mediated NID Syndromes

The retrovirus-mediated NID syndromes are characterized by progressive and premature dysfunction or death of cells in both the CNS and the peripheral immune system. The following is a brief review of NIDs mediated by HIV and *ts*1-MoMuLV infection.

HIV-Mediated NID

It is now well established that HIV infection depletes $CD4^+$ T lymphocytes in the immune system, leading to severe immunosuppression. Although it is not exactly clear how HIV kills $CD4^+$ T cells, a prevalent theory is that the mechanisms by which HIV induces T cell death are multifactorial.[9] In most cases, HIV infection of immune cells triggers a robust defense response marked by increased secretion of cytokines such as TNFα, IL-1, IL-6, and IFN-γ. Although in general the mechanisms by which these cytokines affect HIV replication and cell losses are still unclear, some light has been shed by studying TNFα, the most potent of the HIV-inducing cytokines. TNFα activates the cellular transcription factor NFκB, which induces the expression of a number of cellular genes.[10] The activated cell in turn elevates HIV replication, probably because HIV replicates more efficiently in activated cells. In addition, by binding to the long terminal repeat (LTR) of HIV, NFκB may directly induce HIV transcription,[11] further increasing the rate of HIV replication.

Besides killing T cells, HIV can also infect and activate macrophages and dendritic cells. The activated macrophages may also contribute to the T cell killing and in addition also transport HIV into the CNS.[8,9]

HIV infection also frequently results in neurological disorders.[8,12-14] Although HIV enters the brain quickly in AIDS patients, the majority of patients (except for those with pediatric AIDS) do not experience CNS symptoms until they have become deficient in $CD4^+$ suppressor cells. Neurological symptoms associated with HIV infection include motor, cognitive, and behavioral disturbances.[8] The most serious neurological symptom induced by HIV is profound dementia.[13] Prominent pathological features of AIDS dementia complex (ADC) include neuronal loss,[8,15,16] neuronal apoptosis,[17,18] gliosis, diffuse myelin pallor, and in some cases vacuolar myelopathy.[8] The major target cells for HIV infection in the CNS are macrophages and microglia.[8] Astrocytes are also infected but only at low levels.[13,19-21] Neurons are not directly infected by HIV, suggesting that neuronal loss is caused by indirect mechanisms.

The events leading to neuronal loss seen in ADC are not well understood. However, several factors have been proposed to play a role in neurodegeneration mediated by HIV infection. Several proinflammatory cytokines, TNFα in particular, have been implicated as etiologic factors in HIV-mediated neurodegeneration. ADC is more often correlated with increased levels of TNF than with viral replication in the CNS.[14] Other factors that have been implicated in CNS injury caused by HIV include oxygen-free radicals, nitric oxide, quinolinic acid, and other excitatory amino acid analogs.[13] Recently, dysregulation of signal transduction pathways in astrocytes[22] and astrocyte apoptosis induced by HIV transactivation of the *c-kit* proto-oncogene have been suggested to play a role in HIV-mediated neurodegeneration.[23] Because astrocytes are known to provide neurotrophic support, protect against excitatory amino acid neurotoxicity, and maintain normal ionic and redox homeostasis for neurons,[3,5] it is likely that the dysregulated or apoptotic astrocytes may play a crucial role in the neuronal demise seen in ADC patients.[5,17,24,25]

It is noteworthy that clinically significant motor and cognitive dysfunction to some extent resembling ADC is seen in several animal models, including rhesus macaques infected with certain strains of SIV,[26] cats infected with FIV,[26] and mice infected with *ts1* MoMuLV.[8] The NID syndrome induced by *ts1* MoMuLV is briefly described below. A more detailed description is presented in chapter 4.

tsı *MoMuLV-mediated NID*

The murine retrovirus *tsı* MoMuLV model of neonatal NID characterized by early immunodeficiency, neurodegeneration, and severe wasting resembles in several ways the syndrome produced in humans by HIV, particularly in pediatric HIV infection. Like HIV, *tsı* MoMuLV rapidly depletes $CD4^+$ T cells in the immune system and neurons in the CNS.[27] The destruction of $CD4^+$ T cells and thymocytes by *tsı* MoMuLV is mediated via mitogen-induced apoptosis with nuclear disintegration and membrane blebbing.[28,29] Macrophages are also infected but are not killed by *tsı* MoMuLV.[30] B cell lymphoproliferation and hypergammaglobulinemia are also observed early in *tsı* MoMuLV infections.[27] In the CNS, the prominent pathological features are neuronal loss, neuronal apoptosis, astrogliosis, astrocytic vacuolation, demyelination, and spongiform lesions without peripheral inflammatory cell infiltration.[31,32] Endothelial cells, microglia, and astrocytes, but not neurons, are the primary cells infected by *tsı* MoMuLV. Although neurons are not infected, they are the cells most conspicuously damaged.[8,31,32] Thus, as in HIV, *tsı* MoMuLV appears to mediate neuronal losses indirectly. Astrocytic damage, with production of the death signals TNF, FasL and Fas by astrocytes in areas of the CNS undergoing neuronal death, is sharply increased as the disease progresses,[33] whereas T cell suppressor activity for the activated astroglia (e.g., IL-4) is not detectable in the CNS of *tsı*-infected mice.[33] Therefore, in contrast to T cell death, the indirect neuronal deaths may be partially caused by production of high levels of TNF, Fas and FasL by the activated astrocytes. Alternatively, indirect neuronal death may be caused by failure of the infected astrocytes to properly support the neurons, either by not providing nutrients, redox support, and growth factors or by not controlling excitotoxic amino acid concentrations at neuronal synapses.[3,4,7,33]

NID Syndromes Associated with Genetic Defects

Ataxia Telangiectasia and "Wasted" Mouse

Ataxia-telangiectasia[34,35] and the "wasted" mouse[36,37] model represent familial NID and wasting syndromes in which death of neurons, hematopoietic cells and germline cells is accelerated during postnatal development. The postnatal NID syndrome of AT is caused by mutations in the human nuclear protein kinase gene (*ATM*) or knockout (KO) of the same gene in mice (*ATm*). ATM or ATm modulates cell cycling activity via p53 and NFκB.[35,38-43] When the *ATM* gene is defective, postnatal development of those cells most dependent on external signals for their survival and maturation (e. g., neurons, glial cells, lymphoid cells, and germ line cells) is disrupted. Lymphoid cells, neurons, and germ cells may die prematurely, due largely to failure to balance their proliferative pathways with their apoptotic pathways during postnatal development. Because each of the above cell types develops postnatally at different rates, depending on external signals received during postnatal differentiation, the rates of cell loss during postnatal differentiation vary greatly. In human AT, the debilitating symptoms can vary from immune deficiency to neuronal/astrocytic deficiencies, to germ cell deficiency, to failure of intestinal cell development, and to lymphoid transformation.[3,5,34] In the ATm KO mouse, fetal thymic neoplasia appears at 4 months of age along with impaired postnatal growth and development.[39,44] Human AT lymphoid cells, which have been shown to be unable to produce adequate amounts of their survival factors under mitogenic stress or radiation stress, die prematurely both in vivo and in vitro.[45] In the ATm KO mice, functional and developmental defects in neurons or astrocytes or both are detectable by 2-3 months of age, but neither neuronal losses or astrocytic changes are apparent.[39,41]

Although it is still not clear exactly how the ATM kinase modulates the many nuclear functions required at various times during postnatal development, it appears that the ATM kinase exerts some control over both the major stress-activated transcription factor NFκB and the major tumor suppressor factor p53 in the nucleus.[40,42,46] The NFκB inhibitor IκBα is phosphorylated by ATm,[43] and this leads to proteosomal-induced proteolysis of IκBα, with release of active NFκB.[42,43,47] The timing of expression of the many activities of p53 is modulated by ATm.[40] ATm may also exert direct control of the cell-cycling rate by modulating the activity of the nuclear kinase, Cdc25.[46]

Both lymphoid cells and fibroblasts isolated from ATM-defective humans[35] or ATm-deficient mice[39] are also unable to survive under mitogenic or radiant stress. Differentiation and survival surface markers, e.g., CD3 in ATm-deficient lymphocytes, are also downregulated.[39] In vitro production of survival or growth factors (e.g., IL-2, IL-6, IFNγ, and ICAM-1) under stress may also be impaired.[45] The proteolytic apoptotic pathways to nuclear cell death in the ATM-deficient cells are readily activated under growth factor-deficient conditions.[45] Thus, a major function of the ATM kinase appears to be to modulate those inducible transcription factors that specifically control cell fates during postnatal development of cells of the immune/defense systems in CNS and in the periphery. This resultant dysregulation in cell fate control during postnatal development of cells is likely to be responsible secondarily both for the degeneration of neurons and for the loss of the peripheral immune cells. Whether the loss of neurons is due directly to the ATM deficiency in specific neurons or due to ATM deficiency in the glial support cells is not clear.

"Wasted" Mouse

The phenotype of the postnatal homozygous "wasted" mouse is very similar to that seen in ATM-deficient humans. All of the mouse cells appear normal at birth, but by 4-5 weeks of age, simultaneous and rapid losses of lymphoid cells as well as of cerebellar and/or motor neurons occurs, along with severe wasting.[37,48] The wasting in these mice may be caused by faulty development, after weaning, of adult-type intestinal epithelium with subsequent severe malabsorption, as is seen in the NID syndrome produced by impaired proteolytic processing of essential growth factors in the endocytoplasmic reticulum.[49] Cells isolated from the atrophic thymus of the homozygous "wasted" mice were also shown to undergo rapid apoptotic cell death when mitogenically activated under growth factor-deficient conditions (unpublished data). In vivo these undeveloped, "wasted" lymphoid cells express large amounts of IL-2; however, the receptor for IL-2 is not expressed during thymic T cell development, but is expressed in spleen after completion of neonatal development.[37] It appears, therefore, that the defect in development in "wasted" lymphoid cells is also one of timing, due either to faulty expression of signals or to faulty expression of receivers, as occurs during postnatal neuronal development with imbalanced expression of nerve growth factor (NGF) and its specific receptor TRK-A. In this instance, with deficiency of the ligand NGF relative to TRK-A, apoptosis is the result, whereas with deficiency of the receptor TRK-A relative to NGF, tumorigenesis leading to glioblastoma is the result.[50] Thus, although the genes responsible for this murine postnatal wasting syndrome have not been identified, it appears that genes—possibly those like *ATM* or adenosine deaminase[51] or other genes involved in directing cell fates during postnatal development[47]—are responsible for the losses of T cells seen in this NID model.

Preliminary studies on primary lymphoid cells obtained from wasted mice with postnatal NID indicate that in vitro application of the appropriate survival and growth signals, primarily Th1 cytokines and appropriate redox signals, can also prevent the mitogen-induced apoptotic death of "wasted" lymphocytes (unpublished data). Thus, it appears that "wasted" lymphoid cells, like those in AT or in *ts1* MoMuLV-infected mice, are unable to produce adequate amounts of survival ligands when mitogenically overstressed.

In the CNS of the homozygous wasted mouse, prominent vacuolar (spongiform) degeneration with loss of neurons is seen in the spinal cord and to a lesser degree in the brain stem and cerebral cortex and, as in AT, without an influx of peripheral inflammatory cells.[37,48] An elevated expression of TNFα has also been observed by us in the CNS of these wasted mice (unpublished data). This suggests that TNF in the CNS, as in the retroviral infections (see chapter 4) or in transgenic mice (see chapter 6 and 7), may also play a role in the neuronal death seen in the wasted mouse.

The above findings on defective gene-induced NIDs suggest that external ligands or redox agents that can bypass the blocks in postnatal development induced by deficiencies in genes that modulate the function and fate of postnatally developing defense cells are becoming available. Anti-inflammatory signals or ligands that penetrate the blood-brain barrier, e.g., glucocorticoids, pentoxyfillin, *N*-acetyl cysteine, aspirin, resveratrol, NGF and anti-TNF factors may also become clinically useful.[52-55] Nutrient replacement therapy for neurons that are made deficient by the loss of astrocyte-derived supplies may also be beneficial. Such replacement therapy would include feeding (glucose and lactate) and redox support, together with the provision of appropriate neurotrophins, growth factors, macrophage suppressor factors, and neurotransmitters.[4]

Alzheimer's Disease and Other Fibrillary Protein-mediated Syndromes

Unlike the retrovirus and ATM-gene-induced NID syndromes, in which both the peripheral and the central immune defense systems appear to be directly and etiologically involved in the loss of neurons and immune cells, it is primarily the local immune/inflammatory glial system in CNS that is involved in the localized inflammatory responses seen in a number of neurodegenerative diseases, including AD, DS, and PD, and prion-mediated diseases.[56]

In AD, many of the localized lesions containing various fibrillary proteins including neurofilament proteins, microtubule-associated protein (*tau*), amyloid precursor protein (APP), and presenilins peptides are surrounded by highly activated microglia and astrocytes.[57-59] These surrounding cells produce potentially cytotoxic inflammatory products, including TNF, Fas, IL-1, IL-6, and arachidonic acid products, as well as complement components, antiproteases, and large amounts of proteolytic products of APP.[60,61] Yet, it remains unclear whether these inflammatory responses by the glia are beneficial or harmful to the host. The presence of highly activated glial cells only in areas surrounding the damaged neurons suggests that the excessive release of inflammatory chemokines, redox agents, and fibrillary unprocessed peptides from astrocytes or the dying neurons is locally responsible for the neurotoxicity.[62] Dying cells in the inner core of the AD plaques are probably Fas-positive, GFAP-negative astrocytes that are surrounded by a layer of activated microglia with activated GFAP-positive astrocytes banished to the periphery of the plaque. This arrangement suggests that these displaced astrocytes (to the periphery of the plaques) are prevented from supporting the damaged neurons by the infiltrating microglia.[63] Since a major function of astrocytes is to provide, by direct contact, food and reductive support (i.e., lactate and glucose), to the stressed neurons,[4] it is likely that the encased neurons are lacking astrocyte support and become energy, as well as redox, deficient, resulting in oxidative (free-radical) damage and death. Thus, cutoff of the supply of glucose and glycolytic reductive equivalents from astrocytes to the neurons,[4] plus the proximity of the free radical-producing, activated microglia and dying astrocytes, may account for some of the progressive cell death of neurons.

But what is responsible for the formation of the AD plaques? The major suspect is deposition of the fibrillary fragments of APP, primarily amyloid Aβ 1-42, a fragment of APP that is produced under conditions that alter transport or processing of APP in the ER/Golgi apparatus of neurons and astrocytes.[59,61] These neurotoxic fibrillary fragments of APP have been shown to produce free radicals directly,[64] which may then damage both neurons and astrocytes. Furthermore, the Ca^{2+} and excitotoxic amino acid metabolism in the damaged astrocytes may become sufficiently dysregulated to activate cell death pathways in the starving neurons.[4]

If accumulation of these cytotoxic APP fragments in astrocytes and/or neurons is partly responsible for the formation of AD plaques, what mechanisms are responsible for excessive production of these unprocessed and aggregated peptides? Fibrillary interactive proteins of the amyloid family, whose functions are largely structural and cytoskeletal, as well as the presenilin chaperones,[65,66] are found primarily in terminally differentiated neurons and astrocytes. These large amyloid precursor proteins are proteolytically processed in the ER/Golgi system and normally secreted as soluble fragments. But with mutagenic alterations in the APP gene or in genes whose products are involved in the transport or processing of APP,[66-68] self-aggregating and radical-producing fragments such as Aβ 1-42 may be produced in either astrocytes or neurons.[64] Because these fragments can spontaneously aggregate into highly insoluble inflammagenic fibrils that have a very long biological half-life, the accumulation of these secreted fibers in the ER or Golgi apparatus of astrocytes or neurons may then result in congestion in the ER. This may trigger the ER "overload" response, with release of calcium and H_2O_2 and increased synthesis of the ganglioside GD3[69] in the ER, followed by astroglial differentiation and subsequent astrocyte-mediated neuronal death (see chapter 3). Release of these fibrils outside the cell may also attract and activate the local microglia with the subsequent formation of long-lived multicellular AD plaques containing local inflammatory cells. In familial AD, other genes, as well as specific lipoproteins responsible for the transport, processing, secretion, or synthesis of APP have also been found to be defective.[65,66,70]

AD lesions are also found in normal aging. Such nonfamilial cases probably represent multiple environmental or gene-mediated imbalances produced with aging. These may include protease/antiprotease or protein kinase/phosphatase imbalances or mutations in genes that control protein processing in the ER/Golgi apparatus.[61,67] Age-induced defects in energy supply or in redox scavenging[71] may also lead to aberrant processing in the energy-dependent ER, with subsequent ER-induced programmed cell death in neurons and astrocytes. When the highly metabolically active neurons lose astrocytic support with resultant energy and redox losses, neuronal oxidative death is swift.[71] Neuronal losses are also prominent in many of the defective gene-associated NID syndromes in which insoluble fibrillary peptides, such as prions and neurofilaments, accumulate in neurons or astrocytes.[68,70,72,73] But whether the neuronal degeneration seen in these fibrillary peptide-induced syndromes is caused by the direct cytotoxic effect of the peptides on neurons, or indirectly caused by loss of neuronal support by the activated or impaired astrocytes, has not been resolved.

Recent studies indicate that a variant of an immune system gene, *HLA-A2*, is associated with early onset of AD.[74] But whether the HLA-associated damage of neuronal cells is caused by excessive activation of immune/inflammatory activity of the resident immune/defense cells in the CNS is not known. Appreciation of the role that immune/inflammatory mediators may play in AD pathogenesis has been reviewed recently by Rogers and co-workers.[75]

Down Syndrome

In the Down syndrome (DS), many proteins, including APP and superoxide dismutase (SOD) are overproduced.[76] Cardinal features of DS are hypotonia, growth retardation (especially in the CNS), and congenital defects.[76] Early formation in the CNS of AD-like plaques containing APP fragments surrounded by activated astrocytes is often seen in trisomy 21 (DS) individuals. Mice with trisomy 16 also fail to develop properly and die in utero.[76] Thus, although many diverse gene products of chromosome 21, including the APP gene and SOD (or similar genes of murine chromosome 16), may accumulate excessively in DS, it is likely that the accumulation of excess APP and Aβ1-42 in astrocytes and neurons is, as in AD, responsible for the AD-like degenerative plaques seen in the CNS of DS patients. Neuronal death in DS as in AD is also due, at least in part, to loss of redox support. In studies of primary mixed cultures of neurons and astrocytes containing amyloid deposits obtained postmortem from trisomy 21 subjects, neuronal death could be prevented or slowed by addition of redox support, especially *N*-acetyl cysteine.[77] These observations suggest that neuronal cell death pathways can be modulated by redox support, which is normally supplied in vivo by astrocytes. The overproduction of copper in SOD with unbalanced free-radical metabolism in the trisomic brain may also contribute to the loss or dysfunction of various cells in the CNS.[76,78]

Parkinson's Disease

An Ala-to-Thr substitution at position 53 of the precursor protein for α-synuclein, the non-β component of amyloid that is often found in AD plaques, has recently been found in early onset PD.[79] A major effect of the mutation appears to be to promote intracellular aggregation of the fibrillary α-synuclein into filaments. This suggests that accumulation of α-synuclein may promote the neurodegeneration seen in the substantia nigra and other regions of the PD brain. Extracellular deposition of other mutant fibrillary amyloid peptides, such as Aβ 1-42 in AD or prion peptides in Gerstmann-Sträussler-Scheinker or Creutzfeldt-Jakob disease,[72,80] promotes spongiform lesions similar to those seen in PD. The characteristic lesion in PD is the intracellular proteinaceous Lewy body, which contains large amounts of the aggregated peptides of the α-synuclein family.[79] In contrast, the extracellular deposits of fibrillar peptides seen in the core of the AD plaques contain only 10% of these synuclein peptides and 90% of the amyloid precursor (APP) peptides. Although both AD and PD exhibit localized areas of spongiform degeneration that are usually surrounded by reactive or GFAP-negative (dying) astrocytes, it is only the extracellular deposits in AD or in aging that provoke localized contracted lesions that are surrounded by both the highly reactive microglia and astrocytes. Neither AD nor PD promotes a vigorous response by the peripheral immune/inflammatory system.[56,61-63,68,81]

Thus, it appears that defects in post-translational processing that lead to accumulation of insoluble fragments of various membrane proteins, including viral envelope protein, APP, presenilin, *tau* protein, synuclein, and prion protein may account for much of the spongiform degeneration seen in various NID syndromes, including PD. However, the defensive response of proinflammatory support cells in CNS to these peptide-induced neuronal cell losses varies greatly, from little response to a pure astrocytic response to a microglial plus astrocytic response. The response apparently depends on how well these peptides aggregate and in which cell type they accumulate, as well as on their location in the CNS. Whether these aggregated fragments are directly toxic to neurons and astrocytes or whether they activate production of inflammagenic toxins such as free radicals or cytokines that activate astrocytes but kill neurons remains unclear.[82,83]

Because PD spongiform degeneration may be induced by accumulation of unprocessed peptide fragments in astrocytes or neurons and because the brain's response to these foreign peptide fragments is to activate local glial inflammatory defenses, and thereby diminish the astrocytic support functions, it seems appropriate to include PD as an NID syndrome that primarily involves the CNS immune/defense system. Surprisingly, the secondary suppressive response of the peripheral immune defense system against the overactive astrocytic response in the PD CNS appears to be minimal and obviously inadequate to prevent the excessive astrocytic inflammatory response and the subsequent neuronal death seen in PD.

NID Syndromes Involving Genes with Unstable CAG Trinucleotide Repeats

Overexpression and expansion of the CAG/polyglutamine repeat found on various proteins in Huntington's disease (HD) and in other choreiform syndromes also leads to late-onset neurodegenerative syndromes. Overexpression of these CAG trinucleotide repeats on the spinocerebellar ataxia gene *SCA1* in transgenic mice leads to CNS atrophy with loss of selective brain stem and cerebellar neurons.[84] Transgenic mice carrying the 5' end of a human HD gene with CAG repeat expansion also develop an NID syndrome characterized by severe wasting with atrophy of brain, gonads, and thymus. There are, however, no localized areas of neurodegenerative plaques, gliosis, or inflammatory cell influx associated with CNS atrophy in these mice, but the mice do develop ataxia, limb dyskinesia, involuntary movements, constant tremor, and handling-induced mild epileptic seizures.[85] It appears, therefore, that the cells in the atrophic organs with overexpressed CAG repeats either never develop or are very quickly and effectively phagocytized by the local microglia or astrocytes, leaving no trace of the unprocessed inflammatory peptides or gangliosides in the atrophic brain. Whether these CAG repeats in ER are interfering with the astrocytic support of neurons is not clear. Whether the aggregation and binding of these CAG repeats to essential proteins in the ER with ER clogging is responsible for the cell losses is also not clear. Nevertheless, it is clear that the accumulation, especially in brain, of many foreign or fibrillary proteins, including the CAG repeats and retroviral peptides can lead to postnatal cell loss syndromes, primarily of cells in the neuroimmunogonadal systems. These cell loss syndromes may or may not be accompanied by a reactive inflammatory response by the local immune/defense cells in the CNS. Our interpretation from these studies is that the CAG repeats are directly and mildly neurotoxic but not locally inflammagenic. In contrast, the amyloid-type fibrillary peptides are both neurotoxic and locally inflammagenic. Whether the neuronal losses in the amyloid-type peptides are due to direct neurotoxicity, or to accumulation of toxic inflammatory products, or to both, is not yet established.

Conclusion and Hypothesis

It appears that any condition that interferes with the processing of fibrillary proteins (primarily APP, prion, neurofilaments, CAG repeats, and retrovirus precursor envelope proteins) in CNS cells, and promotes the accumulation of excess fibrillary proteins, may result in either a chronic inflammatory and nonsupportive response by local immune/support cells or simply neuronal atrophy without an overactive local inflammatory response. The outcome depends on whether the fibrillary proteins are directly neurotoxic or whether they excessively activate the local astrocytic defense system with resultant neuronal death.

In the CNS, a prominent characteristic of many NID syndromes is spongiform degeneration. These lesions begin as intracellular vacuoles in astrocytes or in neurons, which may develop into focal areas of large coalescing vacuoles surrounded by cell membranes. Inflammatory reactive cells, primarily astrocytes, are present at the periphery of some of the lesions, while dead or dying neurons and astroglia with various aggregated proteins populate the center.

Apoptotic cell death seen in NID syndromes may represent acceleration of the normal processes of cellular selection and pruning that occur during postnatal development, especially in the immune and nervous systems. This type of apoptotic cell death is accompanied by excretion of phosphatidylserine-coated vesicles. Phagocytotic removal of these vesicles is rapid in those organs having a large phagocytic cell component. In the postnatal thymus and spleen, which contain many phagocytic cells, the complete disappearance of lymphoid cells without spongiform changes is the usual result. In the CNS during postnatal development or in situations of mild neurotoxic insults, the removal of apoptotic cells by the limited number of phagocytes is generally adequate. However, with more severe insults, as in syndromes of amyloid accumulation, the local proimmune/prophagocytic system in the CNS is enlisted. Competition for the available energy in the CNS between the very active neurons and the now activated and proliferating glia may become critical. This may deprive the neurons of the energy needed for their function, resulting in further neuronal losses.

In the perinatal CNS, however, where glial phagocytes are very limited and astrocytes are not yet differentiated into macrophage like cells, indigestible or viral fibrillary membrane proteins may accumulate excessively, especially under conditions of overproduction or underactivity of gene products involved in protein processing or energy production. With continued neuronal death, vacuoles (containing cell membranes, undigested fibrillary proteins, and DNA fragments) are excreted from the dying cells and these may coalesce into large lacunae. In the phagocyte-poor CNS, these lacunae and plaques may persist for a long time.

In the postnatal NID syndromes produced by various genetic defects (e.g., the wobbler mouse,[86] wasted mouse,[37]) the initial lesion seen is early intracellular vacuolization in the CNS. In transgenic mice whose astrocytes are overstressed by overexpression of TNF, IL-3, or IL-6 (see chapters 6 and 7), thereby diminishing the astrocytic support functions for neurons, neuronal cytoplasmic vacuolization is the result.

In adults with HIV infection in the CNS, the vacuolar lesions in the CNS are infiltrated by activated inflammatory cells, primarily microglia and peripheral macrophages. The neuronal losses associated with AIDS dementia complex may, therefore, result either from infiltration of the peripheral activated macrophages with excessive production of neurotoxins, or from lack of astrocytic support due to impairment of these cells as a result of HIV infection in the CNS.

Thus, in the different types of NIDs discussed here, whether caused by retroviruses or by mutated genes, accumulation of various fibrillary and viral proteins may lead to overactivation of glial cells which support neurons. This chronic loss of neuronal support is thought to account for much of the delayed neuronal death seen in these NID syndromes. However, the early responses of astrocytes are initially neuroprotective. But when the damage to astrocytes is severe and the defensive inflammatory maneuvers of the differentiated astrocytes become excessive with overproduction of inflammatory neurotoxins, neurons quickly die. The end result in most cases is spongiform lesions with walled-off, localized plaques containing indigestible fibrillary proteins surrounded by differentiating and/or dying astrocytes and inflammatory cells. Thus, it appears that imbalances in the dual functions of astrocytes, i.e., neuronal support and defense, contribute greatly to the NID syndromes discussed in this chapter.

Acknowledgments

We thank C. McKinley and M. Lynn for their assistance in preparing the manuscript. We also thank Richard Wilcox and Jude Richard for critical reading of the manuscript. This work is supported by Public Health Service grants CA45124 from the National Cancer Institute (NCI), AI28283 from the National Institute of Allergy and Infectious Diseases, MH57181 from the National Institute of Mental Health, Core Grant 16672 from the NCI, and a grant from the AT foundation, Austin, TX.

References

1. Screpanti I, Meco D, Scarpa S et al. Neuromodulatory loop mediated by nerve growth factor and interleukin 6 in thymic stromal cell cultures. Proc Natl Acad Sci USA 1992; 89:3209-3212.
2. Steinman L. Connections between the immune system and the nervous system. Proc Natl Acad Sci USA 1993; 90:7912-7914.
3. Lynn WS, Wong PKY. Neuroimmunodegeneration: Do neurons and T cells utilize common pathways for cell death? FASEB J 1995; 9:1147-1156.
4. Tsacopoulos M, Magestretti PJ. Metabolic coupling between glia and neurons. J Neurosci 1996; 16:877-885.
5. Wong PKY, Lynn WS. Neuroimmunodegeneration. EOS J Immunol Immunopharmacol 1997; 17:30-35.
6. Rothstein JD, Jin L, Dykes-Hoberg M et al. Chronic inhibition of glutamate uptake produces a model of slow neurotoxicity. Proc Natl Acad Sci USA 1993; 90:6591-6595.
7. Lipton SA, Rosenberg PA. Excitatory amino acids as a final common pathway for neurologic disorders. N Engl J Med 1994; 330:613-619.
8. Gonzales-Scarano F, Nathanson N, Wong PKY. Retroviruses and the nervous system, In: Levy JA, ed. The Retroviridae, vol. 4. New York, NY: Plenum Press, 1995:409-490.
9. Fauci AS. Host factors and the pathogenesis of HIV-induced disease. Nature 1996; 384:529-534.
10. Lenardo MJ, Baltimore D. NF-κB: A pleiotropic mediator of inducible and tissue-specific gene control. Cell 1989; 58:227-229.
11. Poli G, Bressler P, Kinter A et al. Interleukin 6 induces human immunodeficiency virus expression in infected monocytic cells alone and in synergy with tumor necrosis factor α by transcriptional and post-transcriptional mechanisms. J Exp Med 1990; 172:151-158.
12. Spector SA, Hsia K, Pratt P. Virologic markers of human immunodeficiency virus type 1 in cerebrospinal fluid. J Infect Dis 1993; 168:68-74.
13. Kolson DL, Pomerantz RJ. AIDS dementia and HIV-1-induced neurotoxicity: Possible pathogenic associations and mechanisms. J Biomed Sci 1996; 3:389-414.
14. Dubois-Dalcq M, Altmeyer R, Chiron M et al. HIV interactions with cells of the nervous system. Current Opinion in Neurobiology 1995; 5:647-655.
15. Ketzler S, Weis S, Haug H et al. Loss of neurons in the frontal cortex in AIDS brains. Acta Neuropathol 1990; 80:92-94.

16. Everall IP, Luthert J, Lantos PL. Neuronal loss in the frontal cortex in HIV infection. Lancet 1991; 337:1119-1121.

17. Shi B, DeGirolami U, He J et al. Apoptosis induced by HIV-1 infection of the central nervous system. J Clin Invest 1996; 98:1979-1990.

18. Espey MG. Apoptosis and HIV neuropathogenesis. Med Hypotheses 1995; 44:536-538.

19. Saito Y, Sharer LR, Epstein LG et al. Overexpression of nef as a marker for restricted HIV-1 infection of astrocytes in postmortem pediatric central nervous tissues. Neurology 1994; 44:474-481.

20. Tornatore C, Chandra R, Berger JR et al. HIV-1 infection of subcortical astrocytes in the pediatric central nervous system. Neurology 1994; 44:481-487.

21. Takahashi K, Wesselingh SL, Griffin DE et al. Localization of HIV-1 in human brain using polymerase chain reaction/in situ hybridization and immunocytochemistry. Ann Neurol 1996; 39:705-711.

22. Wyss-Coray T, Masliah E, Toggas SM et al. Dysregulation of signal transduction pathways as a potential mechanism of nervous system alterations in HIV-1 gp120 transgenic mice and humans with HIV-1 encephalitis. J Clin Invest 1996; 97:789-798.

23. He J, DeCastro CM, Vandenbark GR et al. Astrocyte apoptosis induced by HIV-1 transactivation of the c-*kit* protooncogene. Proc Natl Acad Sci 1997; 94:3954-3959.

24. Price R, Brew B. The AIDS dementia complex. J Infect Dis 1988; 158:1079-1083.

25. Petito CK, Roberts B. Evidence of apoptotic cell death in HIV encephalitis. Am J Pathol 1995; 146:1121-1130.

26. Vitkovic L, Stover E, Koslow SH. Animal models recapitulate aspects of HIV/CNS disease. AIDS Res and Human Retroviruses 1995; 11:753-759.

27. Wong PKY. Moloney murine leukemia virus temperature-sensitive mutants: A model for retrovirus-induced neurologic disorders. Curr Top Microbiol Immunol 1990; 160:29-60.

28. Lynn WS, Wong PKY. Neuroimmunopathogenesis of *ts*1 MoMuLuV viral infection. NeuroImmunoModulation 1998; in press.

29. Saha K, Yuen PH, Wong PKY. Murine retrovirus-induced destruction of CD4$^+$ T cells and thymocytes are mediated through activation-induced death by apoptosis. J Virol 1994; 68:2735-2740.

30. Wong PKY, Prasad G, Hansen J et al. *ts*1, a mutant of Moloney murine leukemia virus-TB, causes both immunodeficiency and neurologic disorders in BALB/c mice. Virology 1989; 170:140-154.

31. Wong PKY, Yuen PH. Cell types in the central nervous system infected by murine retroviruses: Implications for the mechanisms of neurodegeneration. Histol Histopathol 1994; 9:845-848.

32. Stoica G, Illanes O, Tasca S et al. Temporal central and peripheral nervous system changes induced by a paralytogenic mutant of Moloney murine leukemia virus TB. Lab Invest 1993; 66:427-436.

33. Choe WK, Stoica G, Lynn WS et al. Neurodegeneration induced by MoMuLV-*ts1* and increased expression of TNFα and Fas in the central nervous system. Brain Res 1998; 779:1-8.

34. Gatti RA. Ataxia telangiectasia: Genetic studies. In: Gupta S, Griscelli C, eds. New Concepts in Immunodeficiency Diseases. West Sussex, England: Wiley Liss, 1993:202-225.

35. Meyn MS. Ataxia-telangiectasia and cellular responses to DNA damage. Cancer Res 1995; 55:5991-6001.

36. Schultz LD, Sweet HO, Davisson MT et al. "Wasted", a new mutant of the mouse with abnormalities characteristic of ataxia telangiectasia. Nature 1982; 297:402-404.

37. Libertin CR, Ling-Indeck L, Padilla M et al. Cytokine and T-cell subset abnormalities in immunodeficient wasted mice. Mol Immunol 1994; 31:753-759.

38. Savitsky K, Bar-Shira A, Gilad S et al. A single ataxia telangiectasia gene with a product similar to PI-3 kinase. Science 1995; 268:1749-1753.

39. Barlow C, Hirotsune S, Paylor R et al. Atm-deficient mice: A paradigm of ataxia telangiectasia. Cell 1996; 86:159-171.

40. Xu Y, Baltimore D. Dual roles of ATM in the cellular response to radiation and in cell growth control. Genes Dev 1996; 10:2401-2410.

41. Elson A, Wang Y, Daugherty CJ et al. Pleiotropic defects in ataxia telangiectasia protein-deficient mice. Proc Natl Acad Sci USA 1996; 93:13084-13089.

42. Jung M, Zhang Y, Lee S et al. Correction of a radiation sensitivity in ataxia telangiectasia cells by a truncated IκB-α. Science 1995; 268:1619-1621.

43. Jung M, Kondasatyev A, Sung AL et al. ATm gene phosphorylates IκBα. Cancer Res 1997; 57:24-27.

44. Xu Y, Ashley T, Brainerd EE et al. Targeted disruption of ATM leads to growth retardation, chromosomal fragmentation during meiosis, immune defects, and thymic lymphoma. Genes Dev 1996; 10:2411-2422.

45. Lynn WS, Wong PKY. Possible control of cell death pathways in ataxia telangiectasia—a case report. NeuroImmunoModulation 1997; 4:277-284.

46. Furnari B, Rhind N, Russell P. Cdc25 mitotic inducer targeted by Chk1 DNA damage checkpoint kinase. Science 1997; 277:1495-1497.

47. Drexler ACA. Activation of the cell death program by inhibition of proteasome function. Proc Natl Acad Sci USA 1997; 94:855-860.

48. Woloschak GE, Rodriguez M, Krco CJ. Characterization of immunologic and neuropathologic abnormalities in wasted mice. J Immunol 1987; 138:2943-2499.

49. Saftig P, Hetman M, Schmahl W et al. Mice deficient for the lysosomal proteinase cathepsin D exhibit progressive atrophy of the intestinal mucosa and profound destruction of lymphoid cells. EMBO J 1995; 14:3599-3608.

50. Nakagawara A, Nakamura Y, Ikeda Hl et al. High levels of expression and nuclear localization of interleukin-1 β converting enzyme (ICE) and CPP32 in favorable human neuroblastomas. Cancer Res 1997; 57:4578-4584.

51. Beneveniste P, Cohen A. p53 expression is required for thymocyte apoptosis induced by adenosine deaminase deficiency. Proc Natl Acad Sci USA 1995; 92:8373-8377.

52. Grilli M, Pizzi M, Memo M et al. Neuroprotection by asprin and sodium salicylate through blockade of NF-κB activation. Science 1996; 274:1383-1385.

53. Jang M, Cai L, Udeani GO et al. Cancer chemopreventive activity of resveratrol, a natural product derived from grapes. Science 1997; 275:218-225.

54. Henderson CE, Javaheri M, Kopko S et al. Reduction of lower motor neuron degeneration in wobbler mice by N-acetyl-L-cysteine. J Neurosci 1996; 16:7574-7582.

55. Auphan N, DiDonato JA, Rosette C et al. Immunosuppression by glucocorticoids: Inhibition of NFκB activity through induction of IκB synthesis. Science 1995; 270:286-290.

56. Rosenberg RN, Prusiner SB, DiMauro S et al., eds. The Molecular and Genetic Basis of Neurological Disease. 2nd ed. Boston, MA: Butterworth-Heinemann, 1997.

57. Morrison JH, Hof PR. Life and death of neurons in the aging brain. Science 1997; 278:412-419.

58. Morrison-Bogorad M, Weiner MF, Rosenberg RN et al. Alzheimer's disease. In: Rosenberg RN, Prusiner SB, DiMauro S et al.s eds. The Molecular and Genetic Basis of Neurological Disease. 2nd ed. Boston, MA: Butterworth-Heinemann, 1997:581-600.

59. Selkoe DJ. Cellular and molecular biology of the beta-amyloid precursor protein and Alzheimer's disease, *In* Rosenberg RN and Prusiner SB and DiMauro S et al. (ed.), The Molecular and Genetic Basis of Neurological Disease, 2nd ed. Butterworth-Heinemann, Boston, MA 1997:601-611

60. Selkoe DJ. Amyloid β-protein and the genetics of Alzheimer's disease. J Biol Chem 1996; 271:18295-18298.

61. Gabuzda D, Busciglio J, Chen LB et al. Inhibition of energy metabolism alters the processing of amyloid precursor protein and induces a potentially amyloidogenic derivative. J Biol Chem 1993; 269: 13623-13626.

62. Selkoe DJ. Alzheimer's disease: Genotypes, phenotype, and treatments. Science 1997; 275:630-632.

63. Nishimura T, Akiyama H, Yonehara S et al. Fas antigen expression in brains of patients with Alzheimer-type dementia. Brain Res 1995; 695:137-145.

64. Butterfield DA, Hensley K, Harris M et al. β-amyloid peptide free radical fragments initiate synaptosomal lipoperoxidation in a sequence-specific fashion: Implications to Alzheimer's disease. Biochem Biophys Res Commun 1994; 200:710-771.

65. Kim T-W, Pettingell WH, Jung Y-K et al. Alternative cleavage of Alzheimer-associated presenilins during apoptosis by a caspase-3 family protease. Science 1997; 277:373-376.
66. Seeger M, Nordstedt C, Petanceska S et al. Evidence for phosphorylation and oligomeric assembly of presenilin 1. Proc Natl Acad Sci USA 1997; 94:5090-5094.
67. Xu H, Greengard P, Gandy S. Regulated formation of Golgi secretory vesicles containing Alzheimer β-amyloid precursor protein. J Biol Chem 1995; 270:23243-23245.
68. Diedrich JF, Minnigan H, Carp RI et al. Neuropathological changes in scrapie and Alzheimer's disease are associated with increased expression of apolipoprotein E and cathepsin D in astrocytes. J Virol 1991; 65:4759-4768.
69. De Maria R, Lenti L, Malisan F et al. Requirement for GD3 ganglioside in CD95- and ceramide-induced apoptosis. Science 1997; 277:1652-1656.
70. Lee RKK, Araki W, Wurtman RJ. Stimulation of amyloid precursor protein synthesis by adrenergic receptors couled to cAMP formation. Proc Natl Acad Sci USA 1997; 94:5422-5426.
71. Sugrue MM, Shin DY, Lee SW et al. Wild-type p53 triggers a rapid senescence program in human tumor cells lacking functional p53. Proc Natl Acad Sci USA 1997; 94:9648-9653.
72. Hsiao K, Baker HF, Crow TJ et al. Linkage of a prion protein missense variant to Gerstmann-Sträussler syndrome. Nature 1989; 338:342-348.
73. Diedrich JF, Bendheim PE, Kim YS et al. Scrapie-associated prion protein accumulates in astrocytes during scrapie infection. Proc Natl Acad Sci USA 1991; 88:375-379.
74. Payami H, Schellenberg GD, Zareparsi S et al. Evidence for association of HLA-A2 allele with onset age of Alzheimer's disease. Neurology 1997; 49:512-516.
75. Rogers J, Webster S, Lue L-F et al. Inflammation and Alzheimer's disease pathogenesis. Neurobiol of Aging 1996; 17:681-686.
76. Epstein CJ. Down syndrome, *In* Rosenberg RN, Prusiner SB, and DiMauro S et al. (ed.), The Molecular and Genetic Basis of Neurological Disease, 2nd ed. Butterworth-Heinemann, Boston, MA 1997:51-79.
77. Busciglio J, Yankner BA. Apoptosis and increased generation of reactive oxygen species in Down's syndrome neurons in vitro. Nature 1995; 378:776-779.
78. Epstein CJ, Avraham KB, Lovett M et al. Transgenic mice with increased Cu/Zn-superoxide dismutase activity: Animal model of dosage effects in Down syndrome. Proc Natl Acad Sci USA 1987; 84:8044-8048.
79. Polymeropoulos MH, Lavedan C, Leroy E et al. Mutation in the α-synuclein gene identified in families with Parkinson's disease. Science 1997; 276:2045-2047.

80. Ueda K, Fukushima H, Masliah E et al. Molecular cloning of cDNA encoding an unrecognized component of amyloid in Alzheimer disease. Proc Natl Acad Sci USA 1993; 90:11282-11286.

81. Westaway D, DeArmond SJ, Cayetano-Canlas J et al. Degeneration of skeletal muscle, peripheral nerves, and the central nervous system in transgenic mice overexpressing wild-type prion proteins. Cell 1994; 76:117-129.

82. Forloni G, Angeretti N, Chiesa R et al. Neurotoxicity of a prion protein fragment. Nature 1993; 362:543-546.

83. Behl C, Davis JB, Lesley R et al. Hydrogen peroxide mediates amyloid β protein toxicity. Cell 1994; 77:817-827.

84. Burright EN, Clark HB, Servadio A et al. *SCA1* transgenic mice: A model for neurodegeneration caused by an expanded CAG trinucleotide repeat. Cell 1995; 82:937-948.

85. Mangiarini L, Sathasivam K, Seller M et al. Exon 1 of the *HD* gene with an expanded CAG repeat is sufficient to cause a progressive neurological phenotype in transgenic mice. Cell 1996; 87:493-506.

86. Mitsumoto H, Ikeda K, Klinkosz B et al. Arrest of motor neuron disease in *wobbler* mice cotreated with CNTF and BDNF. Science 1994; 265:1107-1110.

CHAPTER 3

Pathogenic Mechanisms in Neuroimmunodegeneration

William S. Lynn and Paul K.Y. Wong

Introduction

The number of signals now known to modulate cell survival and death of neuronal and immune (peripheral and central) cells is increasing rapidly. The timing and concentration of these signals are especially crucial during postnatal development when rapid adaptation to ever-changing environmental signals is required. During this period, cell growth must be carefully balanced with cell death, and cell differentiation must be commensurate with the rapidly changing environmental signals, especially those signals produced by nutrients or by their interacting neighboring cells and matrix. The number and activity of each of the cellular organelles must also be carefully adjusted to meet the specific and rapidly changing needs of the cells. Interactions between the cells and their neighboring cells and matrix must also be exquisitely programmed. During this period of interactive, postnatal development, errors and imbalances are frequent. Under situations of either inborn errors (e.g., genetic defects) or of excessive environmental stresses (e.g., viral infections), unrepaired genetic mistakes and indigestible oxidized peptides accumulate. Cell death, either apoptotic or oxidative, especially in the immune and nervous systems, becomes imminent. In the neuroimmunodegeneration (NID) syndromes outlined in chapter 2, cell death is cell induced (apoptotic), and therefore may not alert the peripheral migratory immune system but usually does alert the CNS immune/support system. The initial response of the CNS immune/support system is to defend and protect the apoptotic neurons. However, the defensive maneuvers of the CNS immune/support cells may

Neuroimmunodegeneration, edited by Paul K.Y. Wong and William S. Lynn.
© 1998 Springer-Verlag and R.G. Landes Company.

become excessive and usurp their neuronal support functions. The loss of neuronal support as a result of impaired function of the CNS immune/support cells may account for much of the neuronal death seen in these NID syndromes.

The purpose of this chapter is to identify the potential pathogenic mechanisms involved in the various NID syndromes described in chapter 2. Because premature cell losses in the immune (central or peripheral) and nervous systems are the cardinal feature in NID syndromes, cellular mechanisms that control the fates of these cells are emphasized here. Potential therapies designed to prevent the losses of these specific cell are discussed. In addition, because most of the neuronal cell losses may be caused by dysfunction of CNS support/defense glial cells, the etiologic role of these defensive responses in NID syndromes is addressed. Potential therapies designed to replace the missing astrocytic support functions are emphasized.

Control of Cell Fate by Organelle-Induced Signals

The fate of cells that must continue to develop and differentiate postnatally in response to their changing environment (e.g., hematopoietic cells, germline cells, intestinal mucosa, astrocytes or neuronal cells) depends largely on the dose and duration of the specific signals they receive during postnatal development. During this time, these cells are very vulnerable to excessive or unbalanced stimuli or stress. Major responses of cells to these stimuli are proliferation, differentiation, transformation, cell death, or senescence. These cellular responses are usually initiated by activation or damage to one of the following cellular organelles: the plasma membrane, the mitochondria, the endoplasmic reticulum (ER)/Golgi apparatus, the actin cytoskeleton or the nucleus (Fig. 3.1). Furthermore, it is now recognized that each intracellular organelle can also independently generate multipurpose signals that activate pathways in other organelles. These signals in concert control the life and death of the cells. The following is a brief review of the organelle-produced signals that control cell fate.

Plasma Membrane

A major control for cell fate at the plasma membrane involves receptor-ligand binding. At the plasma membrane, binding of ligands such as FasL, TNF, NGF and c-KitL to their respective receptors can activate either cell death or cell growth signaling pathways, depending largely on the concentration of the ligands. All of these receptors

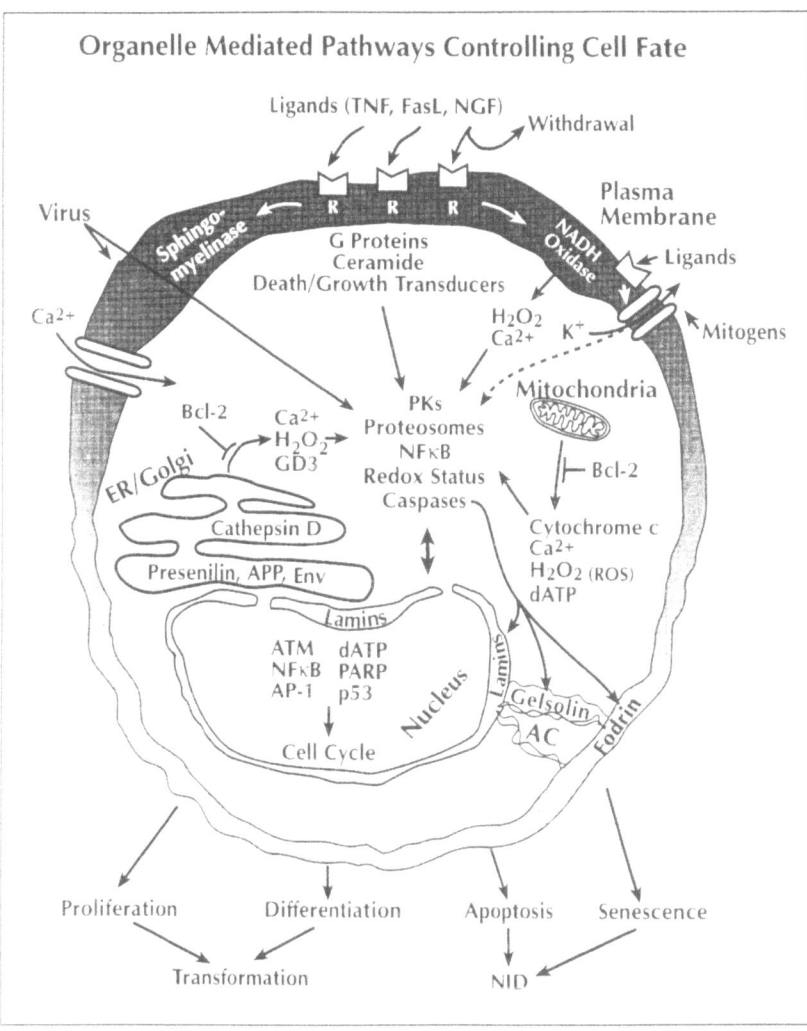

Fig. 3.1. Schematic presentation of possible mechanisms of cellular organelle-mediated cell fate. The fate of cells depends largely on the dose and intensity of specific signals they receive. Major responses of cells to these signals are proliferation, differentiation, transformation, apoptosis, or senescence. These cellular responses are usually initiated by activation or damage to one of the following cellular organelles: the plasma membrane, the mitochondria, the endoplasmic reticulum/Golgi apparatus, the nucleus and the actin cytoskeleton. Each of these organelles can not only independently generate multipurpose signals that control cell fate, but also activate other pathways that control cell fate in other organelles. Abbreviations: R, receptor; PKs, protein kinases; APP, amyloid precursor protein, ATM, AT protein kinase; Env, viral envelope protein; GD3, ganglioside; PARP, poly(ADP-ribose)polymerase; AC, actin cytoskeleton, ROS, reactive oxygen species.

(except Fas) are coupled to both pathways. At high ligand concentrations, binding of ligand to receptor may activate membrane Ca^{2+}-ATPases, NADH oxidases,[1] sphingomyelinase,[2,3] phospholipases, lipoxygenases,[4] or proteases,[2] which may then release large amounts of Ca^{2+}, H_2O_2 and ceramides into the cell and cause the release of K^+ from the cell.[5] If not properly controlled, these activities will lead to cell shrinkage and death. At lower or balanced concentrations of ligands and receptors, the signal from the receptor-ligand complex causes cell proliferation and suppresses apoptotic cell death. Examples of this type of receptor-ligand interaction include c-Kit in hemopoietic cells,[6] TNF in adipocytes,[7] and NGF in neurons.[8] Severe imbalances in the ratio of c-Kit to its ligand at the plasma membrane of astrocytes also lead to apoptosis.[9] Similarly, HIV Nef protein induces apoptosis in astrocytes by excessive transactivation of the c-Kit promoter.[9] Retrovirus products, e.g., envelope protein or its fragments, may promote apoptosis by directly activating death pathways at the plasma membrane via cAMP.[10] Lower concentrations of the envelope proteins are mitogenic[11,12] but whether these mitogenic viral peptides are activating proliferative pathways via signaling by the voltage gated K^+ channels at the plasma membrane[13] is not clear. The hydrophobic peptides derived from the envelope protein of these retroviruses may also, like daunorubicin and other hydrophobic ketones,[2] activate thiol-mediated proteases at the cell surface, which in turn activate neutral sphingomyelinase and liberate ceramides that initiate the mitochondrial and caspase-dependent pathways to apoptosis.[2,3,14]

Another major control for cell survival and apoptosis at the plasma membrane is the expression of the inducible glucose transporters Glut-1 and Glut-4.[15] Upregulation of these transporters by cytokine growth factors with transport of these transporters to the cell surface leads to proliferation. Downregulation, as by growth-factor withdrawal or deficiencies in the protein tyrosine kinases, Abelson protein kinases, or the insulin receptor, leads to energy-dependent apoptosis.[15-17] Alternate energy sources, such as pyruvic acid or glutamine, which can bypass the glucose transporter-dependent energy deficiency, may be useful in preventing this type of plasma membrane-mediated cell death. Whether the hydrophobic fibrillary proteins, or viral envelope proteins, can inhibit the function of these glucose transporters is not known. Activated T cells, which normally transport glucose rapidly in culture and homotypically aggregate rapidly when exposed to mitogens, lose these two functions in parallel when they are deprived of growth factors or infected by *ts1* MoMuLV[11] or when their Ras or other receptors are ligated.[16,17] The Ras-dependent loss of Glut-1 transporter function in such cases, which is caused by decreased affinity for glucose, is likely secondary to c-abl-induced alteration in tyrosine phosphorylation of the transporter.[15,17] Although the mechanism involved in the homotypic aggregation and glucose transport are not well understood, the close temporal association of the ability of cells to aggregate and transport glucose rapidly suggests that ligand-induced conformational constraints at the plasma membrane which alter rates of K^+ efflux may be involved in activation of apoptotic signaling pathways. However, whether binding of viral envelope or other fibrillary proteins at the cell surface can alter membrane conformation and thereby downregulate glucose and K^+ transport and homotypic aggregation is yet to be evaluated.

Role of Cell-Surface Thiols and Proteases

Cell-surface proteases and thiols are now known to be involved in activating apoptotic cell death pathways. Activation of sphingomyelinase at the cell surface by TNF or by hydrophobic ketones, such as daunorubicin or dichloroisocoumarin, with liberation of the potent apoptosis-inducing ceramides, is mediated by these proteases.[2,3,18] These proteases are blocked by serine protease inhibitors but not by inhibitors such as leupeptin or pepstatin A, which block endosomal aspartyl proteases (e.g., cathepsin D). Although this cell surface proteolytic activity has not been well characterized, the fact that it is greatly modulated by hydrophobic complex ketonic nitrogenous compounds (e.g., daunorubicin, coumarins, and chloromethyl ketones) suggests that thiol or redox interactions at the plasma membrane may induce the proteolytic activity responsible for activation of the neutral sphingomyelinase and the subsequent serine protease-induced (AP24) DNA fragmentation.[3] Interactions of the ketonic protease inhibitors[2] with the exposed surface thiols or the protein disulfide-isomerase thiols at the plasma membrane[19,20] may also be involved in exposing the hydrolytic activity responsible for activation of the sphingomyelinase.[2]

Addition of impermanent thiols or oxidants has been shown to modulate rates of cell death and survival both positively and negatively,[11,21] but whether these surface thiols[20] are involved in mediating the changes in cell growth or apoptosis induced by retroviral envelope proteins is not clear.

The superfamily of transmembrane tetraspanins that stabilize interactions of dimerizing proteins in the plasma membrane[22] are also involved in modulating the many diverse functions controlled at the plasma membrane, including cell activation, proliferation, motility, adhesion, differentiation, Ca^{2+} uptake, K^+ efflux and apoptosis.

Mitochondria

Mitochondria produce many growth/death signals such as H_2O_2, Ca^{2+}, dATP,[23,24] and cytochrome c,[24-26] as well as energy (via ATP). They use these products as signals for activation of pathways to either cell death or survival and proliferation. Mitochondrial ATP together with redox support from glucose metabolism is required for optimal cell death and cell growth.[27] Glycolytic redox support is also required for proteolytic cell death programs.[28] Release of cytochrome c from the outer mitochondrial matrix into cytoplasm is primarily a signal for activation of the caspase cascade to cell death,[25,26,29] whereas release of Ca^{2+} or H_2O_2 in low, but not high, amounts is a signal for cell growth.[30] Liberation of these factors from mitochondria is controlled by the Bcl-2/Bax family of inducible antiapoptotic peptides, which by binding to mitochondrial outer membrane pores[29,31] stabilize the membrane pores against leaks and consequently block cytochrome c release and apoptotic signaling. The Bcl-2 family may also function in other intracellular membranes as "molecular facilitators," much like the tetraspanins.[22] Because expression of both the tetraspanin family and the Bcl-2 family is inducible primarily in the immune and nervous systems,[22,29] it is likely that both of these membrane-stabilizing systems are involved in modulating the mitochondria-induced apoptosis in the NID syndromes discussed here.

Damage to mitochondrial membranes directly by uncoupling or indirectly by signals from plasma membrane (e.g., ceramide),[14,32] also leads to release of Ca^{2+}, H_2O_2, and cytochrome c from mitochondria[24,25] and subsequently to activation of the Ca^{2+} and cytochrome c-mediated cell death pathways. Chronic activation of G-protein-coupled phospholipases[33] at the plasma membrane by TNF also leads to dissolution of mitochondrial cristae, with uncoupling and production of reactive oxygen molecules, and finally to inactivation of electron transport at complex III.[34] Because retroviral infection of astrocytes in cultures or in the CNS also leads to excessive production of TNF, it is likely that the observed neuronal degeneration[35,36] is caused in part by the indirect activation of the TNF-mediated mitochondrial pathways to cell death.

Endoplasmic Reticulum/Golgi Apparatus

Role of Misfolded or Aggregated Fibrillary Peptides

There is growing evidence that abnormal protein folding and trafficking with resultant accumulation of aggregated proteins in the ER/Golgi apparatus may lead to ER-induced disease.[37,38] For example, scrapie-associated prion protein accumulates in astrocytes during scrapie infection[39] and increased aggregation of Ab peptides is seen in neural cells in AD patients.[40] In *ts1* MoMuLV and HIV-mediated NIDs, accumulation of unprocessed viral envelope or prion proteins in the ER is seen.[41] Whether the abnormal protein folding and accumulation in ER actually causes apoptosis, however, is unclear.

There are several possibilities for how protein misfolding with accumulation in the ER can lead to cell death. The first possibility is that the accumulation of aggregated viral and cellular membrane proteins in the ER may promote release of Ca^{2+} from ER, which in turn activates the transcription factors AP-1 and NFκB.[42] Although Ca^{2+}-dependent activation of transcription is usually a cell defense mechanism, cell death may also be accelerated if the Ca^{2+} concentration remains chronically elevated. The second possibility is that the traffic of vital cellular receptors and growth factors to the plasma membrane may be blocked in the viral peptide-clogged ER. In time, this would lead to growth factor-deficient cell death. The third possibility is that continued protein accumulation may cause ER to swell and release large amounts of H_2O_2, thereby activating the redox-sensitive apoptotic pathways. The fourth possibility is that the accumulated foreign proteins may block the production or activity of chaperones, such as immunoglobulin heavy-chain binding protein (BiP) or protein disulfide isomerase (PDI), that facilitate protein folding and trafficking in the ER.[37,43] The fifth possibility is that the ER-derived ganglioside GD3, which promotes mitochondrial outer membrane leaks,[44] may lead to excessive release of cytochrome c, H_2O_2, and Ca^{2+} from the mitochondria, with induction of caspase or redox-mediated pathways to cell death.[21,29,45,46]

Potential Role of Cathepsin D

The endosomal/lysosomal aspartyl protease cathepsin D (CTSD) is required for cell survival during postweaning development, primarily in those tissues that must rapidly grow or mature postnatally (i.e., lymphoid cells, gastrointestinal epithelia, astrocytes, and neuronal cells). Knockout of the CTSD gene by gene targeting in mice induces, at 2-3 weeks of age, a syndrome of sudden-onset wasting with rapid losses of lymphoid cells and intestinal villae.[47] These homozygous knockout mice are normal preweaning, but die of inanition and apoptotic cell death by 3-4 weeks of age. A primary function of CTSD is to process those inactive precursor peptides (containing hydrophobic aggregating domains) that serve, when processed, as essential signals for the animal to meet the immune, digestive, and behavioral stresses required during postweaning development. Without these fully processed growth factors, onset of apoptotic cell death in the above organs is rapid and fatal. Thus, the specific processing and cleavage in endosomes of transmembrane growth and survival factors required during postnatal development appears to be a major function of activated CTSD. For example, the infection of T cells and astrocytes by *ts1* MoMuLV, which results in accumulation of nonprocessed precursor envelope peptides in the ER,[48] may lead either to competitive inhibition by the virus envelope of the normal processing of essential growth factors in ER or to direct inhibition of the processing and activation of CTSD itself.[47] In either case, the resultant CTSD deficiency with a lack of growth factors would lead to loss or dysfunction of T cells and astrocytes during neonatal development. The result would be wasting, malnutrition, immune deficiency, and death of those neurons (e.g., motor or cerebellar neurons) that must rapidly develop postnatally and are most dependent on their astroglial support cells during this period of differentiation.

Alternatively, overexpression or premature activation of CTSD by the accumulated viral peptides (as may also occur with the death signals Fas, TNF, or IFNγ), may lead to proteolytic activation of other proteases, e.g., caspases, involved in the cell death pathways.[4,45,49] Thus, either downregulation or upregulation of the major protease responsible for endosomal processing during neonatal development of peptides with hydrophobic domains may be pivotal in controlling the degenerative cell losses seen in viral-induced NID syndromes. Because the wasting NID syndrome induced by the *ts1* envelope is both temporally and clinically very similar to that seen in CTSD deficiency, it is tempting to conclude that CTSD deficiency may be involved in the retrovirus-induced syndromes.

In studies using protease inhibitors specific for either CSTD or the caspases, two types of degenerative (apoptotic) changes have been observed.[49] In cycling U937 cells undergoing cell death induced by TNFα or Fas, intracellular nuclear condensation with fragmentation, as well as disruption with vesiculation at the plasma membrane, is seen by 5-6 hours. Treatment of this TNF-induced cell death with the specific aspartyl proteinase inhibitor pepstatin A blocks the disruption and vesiculation reaction at the plasma membrane, but not the intracellular chromatin condensation and fragmentation.[49] Activation of CSTD is therefore required for the vesiculation reaction at the plasma membrane but not for the nuclear destruction. Furthermore, specific peptidic inhibitors for the caspase protease family completely prevented the cytokine-induced nuclear destruction, but not the vesicular response, at the plasma membrane.[49] Thus, although activation of either protease pathway leads in time to apoptotic cell death, it appears that CTSD is required for the vesiculation reaction at the plasma membrane, whereas the caspases are responsible for destroying chromosomal integrity and structure, probably by entering the nucleus with activation of poly(ADP-ribose) polymerase (PARP), and laminin, with subsequent fragmentation of DNA.[46] Because inhibitors of either of these protease families can greatly slow the rate of cell death, both protease families are apparently required for early induction of rapid cell death. But whether the inefficient processing of *ts1* envelope peptides in the ER,[50-52] with accumulation of envelope peptides in the ER/Golgi apparatus, is the signal that leads to activation of either of these two protease-induced pathways to cell death has not been established.

Nucleus

In the nucleus, the accumulation of double-strand breaks in DNA due to irradiation, to Ara-C-induced defects in DNA recombination or repair rates or to dysregulated nuclear tyrosine kinases, i.e., ATm, DNA-pK, or c-Abl,[53-56] leads to activation of nuclear cell death pathways. The substrates for these nuclear kinases, including p53, c-Abl, and RNA polymerase II,[54] are involved in modulation of Gl-S and G2-M checkpoints as well as in modulation of cell death pathways. But whether these apoptotic pathways initiated in the nucleus are influenced by alterations in cytoplasmic concentrations of Ca^{2+}, K^+ and H_2O_2 induced by damage to mitochondria, plasma membrane or ER is not clear. A loss of energy caused by excessive consumption of ATP, either by excessive DNA repair with excessive activation of PARP or by accumulation of dATP in the nucleus,[24,27] could also initiate the apoptotic nuclear destruction. Mutations in genes (e.g., p53 or ATM) that control cell cycling as well as cell death pathways have also been shown to upregulate nuclear pathways to chromatin destruction.[57-59] The loss of K^+ from the cell, with resultant shrinkage of the cell as well as the nucleus, as occurs with excessive or premature mitogenic activation,[5,13] may also upregulate the apoptotic destruction of the nucleus.

Whether the retroviruses can directly activate these nuclear pathways to cell death is not clear. However, in tumor cells overexpressing various viral oncogenes, a senescence or aging type of oxidative cell death is rapidly induced by overexpression of genes that control cell-cycling rates (e.g., p53 or p21).[59] In this type of cell death, the cell remains metabolically active but flattens out and accumulates massive amounts of lipofuscin granules. The mitochondria and nucleus in these senescing cells swell rather than shrink, but their DNA is not fragmented, nor are apoptotic vesicles excreted.[59,60] The risk of transformation in these senescing cells, as in cells infected with these mitogenic viruses, is also greatly increased. This type of nucleus-induced oxidative cell arrest and death is likely occurring, to some extent, in both the gene-induced (ATM) and retrovirus-induced (*ts1* MoMuLV) NID syndromes. This is because in both types of NID, the balance between the metabolic energy (oxidative and reductive) input and the nuclear p53-mediated output (cell cycling and DNA repair) is precarious and may lead to either cell death or proliferation.[28,60] Whether lipofuscin or the aging indicator neutral β-galactosidase accumulates in these nuclear-induced NID syndromes has, however, not been evaluated.[59]

Actin Cytoskeleton

The major pathway to cell death in the cytoplasm is mediated by the family of aspartate-specific caspases which when activated can destroy both the nucleus and the cytoskeleton. Following transport of these activated caspases into the nucleus, nuclear lamin is hydrolyzed and PARP is cleaved.[46] Nucleosomal endonucleases are also activated with resultant fragmentation of DNA.[24,27,45] The cytoskeleton and the cell shape are also destroyed, at least in part, by the proteolytic activity of caspase 3 on gelsolin.[61] The resulting amino terminal cleavage product of gelsolin causes rapid calcium-independent depolymerization of the actin cytoskeleton with loss of cell shape and matrix connections, as well as increased blebbing at the plasma membrane and DNA fragmentation. Whether the gelsolin-induced destruction of the cytoskeletal links to extracellular matrix and plasma membrane is directly responsible for activating the nuclear cell death pathways is not established. Since disruptions in extracellular matrix are a major initiator of apoptotic pathways in some cells, it is likely that cytoskeletal rupture can also activate apoptotic pathways. Fodrin, a major component of the plasma membrane-associated cytoskeleton, is also cleaved by the activated caspase family.[62] Thus, the caspase family of proteases, when activated, rapidly dismantles cytoskeleton components both intracellularly and at the plasma membrane. This dismantling is apparently responsible in part for the blebbing and vesiculation seen in cells undergoing cytoplasmic apoptosis.[27,62]

Etiologic Role of Defensive Responses in NID Syndromes

Many injuries in the brain (e.g., those caused by anoxia, viral infection, excess inflammatory cytokines, glucose deficiency, oxidant stress, inflammagenic peptides or lack of survival signals) that cause rapid damage to neurons and oligodendrocytes result in activation, differentiation, and/or proliferation of those undifferentiated astrocytes whose major function is to support and defend neurons. When the injury is mild, local, and insufficient to rapidly activate neuronal cell death, the slow accumulation of broken or oxidized DNA, lipofuscin or indigestible fibrillary proteins (amyloid, synuclein or prion peptides) may occur. Because neurons are terminally differentiated but actively metabolizing and have lost their ability to adapt defensively to injuries, the continuous accumulation of oxidants, fibrillary proteins and mutations results in a slow, oxidative death. But in the support cells, which maintain their adaptability to stress (i.e., their proliferative or differentiative capacity), this chronic injury with accumulation of oxidants, fibrillary proteins, and cell debris leads initially to defensive activation and differentiation and in time to proliferation. With chronic injury, e.g., prolonged mitogenic signaling with the accumulation of cytoplasmic Ca^{2+}, H_2O_2, cytochrome c, dATP, and ceramides and with increased efflux of K^+, both apoptotic and necrotic pathways to cell death may become activated.[13,62] Both Na^+ influx with cell swelling (necrosis) and K^+ efflux with cell shrinkage (apoptosis) ensue, and the demand for more energy is greatly increased.[5] With the ensuing energy deficiency, excitotoxic neu-rotransmitters may accumulate in the CNS with further loss of in-tracellular K^+ and elevation of H_2O_2, intracellular calcium, GD3 gan-glioside, dATP and cytochrome c. Activation of redox-sensitive defensive pathways by signals that can induce either transcription factors (NFκB and AP-1), proteases (caspases or cathepsin D), or antioxidants (including the Bcl-2 family,[29] and glutathione peroxi-dase) also may occur all dependent on the cell type. Proapoptotic factors, e.g., cytochrome c and iron from mitochondria,[26] GD3 gan-glioside from the Golgi apparatus,[44] and arachidonic acid and its products from membranes, may be released. Other defensive ma-neuvers by support cells, which may become overactivated, include the following signals:

Membrane NADPH and NADH Oxidases

The O_2^- and H_2O_2 produced by the activated membrane en-zymes NADPH and NADH oxidase[1,63] upregulate cellular defenses via cytoskeletal cross-linking and expression of adhesins. However, in excess, these oxidizing signals may become potent cytotoxins, which can be controlled by addition of external reductants.[64-67]

Nitric Oxide (NO) Synthases

These inducible heme enzymes produce the vasodilator NO rather than O_2^- and H_2O_2. Neurons that can be induced to produce NO are scattered in small islands throughout the brain in associa-tion with other neurons that produce endocrine signals, e.g., soma-tostatin, neuropeptide Y, or α-aminobutyrate. NO is also produced in large amounts in the brain by microglia, astrocytes, and endothe-lia. At low doses, NO, like H_2O_2, is neuroprotective, depending on the intracellular redox state.[66] However, higher doses of NO are highly toxic for neurons but not for astrocytes.[66,67] When in excess, NO may combine with O_2^- to produce the more toxic radicals peroxynitrite (ONOO⁻) and hydroxyl radical (OH), with resultant membrane ni-tration and oxidation. Influx of Ca^{2+}, efflux of K^+, outpouring of excitotoxic amino acids from astrocytes, and activation of Ca^{2+}-dependent intracellular proteases, lipases, and nucleases soon follows.[64] These potent products of NO and O_2^- can also inactivate many enzymes, including mitochondrial NADH dehydrogenases, succinate dehydrogenase, glyceraldehyde-3-phosphate dehydroge-nase, and cytochrome oxidases, and can cause release of the potent apoptosis signals cytochrome c and Ca^{2+} from mitochondria.[25,26] NO can also inactivate NMDA receptors as well as cause double-strand breaks in DNA.[64] Thus, in the presence of excess NO, neurons are rapidly deprived of energy, and their macromolecules become oxidized and nitrated. With the continuing damage, mitochondrial, nuclear, and plasma membranes become leaky, the cytoskeleton is disrupted and the inner plasma membrane phospholipids (i.e., phosphatidyl serine) are exposed. This exposure of phospholipids is a signal to local phagocytes to ingest the dying cells or axons.[68] The local lesions that are produced by excess NO are islands of cellular debris, including indigestible oxidized phospholipids, fibrillary protein fragments (amyloids and/or neurofilaments),[40] and synuclein peptides[69] that are usually surrounded by highly activated or dying astrocytes or microglia.

Thus, it is clear that the local immune/inflammatory defenses in CNS can be either neuroprotective or neurodestructive, depending largely on the dose of the inciting agent and the suppressor capabilities for the overactive CNS defense system (Th2 cytokines and glucocorticoids) of the peripheral immune system. Therefore neuronal death or survival in conditions produced by genetic defects, mitogenic viruses, or energy or redox deficiencies, ultimately depends on the ability of the host cell to maintain its balance between its defenses and survival signals. When the injury is chronic or lifelong, proper maintenance of defenses over time becomes faulty, and NID syndromes of cell loss result. Maintaining the proper balance between metabolic input and nuclear consumption may also slowly falter over time, leading to irreversible senescent-type cell death.[59]

Role of Multipurpose Signal Transduction Systems in Controlling Cell Fate

Defense cells, including lymphoid cells and dendritic cells in the thymus, and inducible macrophagelike cells such as astrocytes in the CNS, are highly endowed with multipurpose receptor-mediated signaling systems. These signaling systems can activate defense systems (i.e., proliferative, survival, or repair pathways) as well as cell death pathways. Large energy inputs are required to maintain both types of response. However, under conditions in which energy is rapidly depleted, the depleted cells, especially the metabolically active ones, quickly swell, lose redox control, oxidize their membranes, and die a necrotic, lytic death.[27] Under less stressful conditions in which some reductive glycolytic energy remains, cellular defense systems, either proliferative or apoptotic, are deployed. The choice of defense depends largely on the availability of reductive energy from glycolysis,[28] the extent and type of damage, and the potency or dose of the dual-purpose signals. For example, with low doses of TNF, Fas ligand, NGF, c-Kit ligand, platelet derived growth factor (PDGF), or mitogenic virus, the defensive response is proliferative. At higher doses, the defensive response is apoptotic, a response in which the damaged cells are converted to membrane-bound vesicles that can be ingested by local phagocytic scavengers without alerting the peripheral inflammatory system. For example, at low doses, the dual-purpose receptor/ligand complex (c-Kit/c-KitL) suppresses apoptosis and promotes proliferation.[6] With over-expression of c-Kit, as induced by the transactivating gp120 envelope peptides of HIV, cell cycle progression becomes uncontrolled, and activation of apoptotic pathways is the cell response.[9] Similarly, PDGF, which in the presence of energy and adequate amounts of cell cycle progression factors is a potent proliferation stimulator, may induce apoptosis and cell death under conditions of growth factor or energy deficiency.[56] As previously discussed, the mitogenic retrovirus *ts1* MoMuLV also induces apoptosis at high viral doses but proliferative responses at lower doses. At low doses of *ts1* MoMuLV infection, the activated astrocytes survive by shutting down their p53-dependent cell-cycling pathways and differentiating into slow-growing virus-resistant cells (unpublished data). But, with the continuous mitogenic drive induced by the virus, plus upregulation of cellular survival factors and downregulation of apoptotic pathways, the risk of oxidative senescence in astrocytes or mutagenic transformation as in infected lymphoid cells may be greatly increased.

Thus, it appears that most growth factors or retrovirus can play multiple roles in controlling cell fates. They can modulate both cell proliferation and cell death pathways, with the final outcome for the cell depending largely on the time-dependent concentration of the growth factors. At high or chronic dosages or with deficiencies in the many counter or suppressor factors for apoptosis, e.g., Bcl-2/Bax,[31] $I_\kappa B$,[70,71] p53,[72] ATM,[57] Vpr,[9] cAMP,[10,73] activation of apoptotic pathways is the result. However, under severe conditions in which reductive energy becomes limiting, e.g., stroke or hypoglycemia, energy- or redox-dependent pathways quickly fail and the NMDA-coupled neurons swell, lose control of their ionic pumps and their redox systems, produce blebs and may eventually lyse.[66] In contrast, defensive cells that are capable of adapting to these stresses, e.g., embryonic myoblasts,[74] fibroblasts, and astrocytes,[35] can survive by differentiating into slow-growing resistant cells whose antioxidant and antiapoptotic defenses are now upregulated.

Thus, most cells can respond to varying doses of growth signals in multiple ways, either by cell survival and growth or by differentiation and transformation or by suicide. The response of short-lived immune cells and excitable neurons is usually suicidal, but the response of the long-lived, adaptable defensive cells is usually adaptive differentiation. The end result in NID syndromes is that the short-lived T cells and/or excitable neurons die prematurely while the long-lived cells (macrophages, fibroblasts, myoblasts, plasma cells, and astrocytes) either survive and proliferate excessively or differentiate into resistant cells (see Fig. 3.1).

Cell-fate Homeostasis in NID: Role of p53

A major means that most mitogenically activated cells use to control their fate is the manipulation of the tumor suppressor gene p53. p53 is a major DNA-binding protein that activates transcription of genes involved in cell-cycle arrest, survival, senescence and apoptosis. By controlling the expression of p53 product, mitogenically activated cells can undergo either premature senescence, apoptosis, or prolonged proliferation.[75] With no p53 activity, continuous cell growth and tumor formation is the fate. At low doses of p53, growth arrest and subsequent oxidative damage is the fate.[59,60] At high doses of p53, rapid apoptosis with nuclear disintegration is the end result.[2,59,60,72] At low doses of p53 but very high doses of proliferative signals, oxidative senescent changes with early cell death may occur rapidly (8-20 hours).[59] With slower and better controlled mitogenic stimulation plus low doses of p53, the cells may survive but proliferate at slower rates, only to slowly release ceramide and the mitochondrially active GD3 ganglioside and induce senescence with oxidative degeneration.[32,44,59] With removal of most of the mitogenic stimulation, mature cells, such as human fibroblasts, or astrocytes can lapse into a quiescent state in which ceramide, cytochrome c, or GD3 ganglioside release as well as p53-mediated activities are suppressed. In this state, the cells can survive indefinitely with minimal metabolic activity and without accumulation of DNA damage or oxidized cell debris.[32,59] Thus, p53 is a major controller of both cellular longevity and degeneration. It is, therefore, not surprising that mutations in p53 and other inhibitors of cell-cycle progression (e.g., p21 and p16^{INK4a})[75] are involved in a large number of human degenerative syndromes and cancers.[76,77] Although the precise role of p53, and its many modulators (including the nuclear protein kinase, ATM[55,78]) have not been fully delineated in the many NID syndromes discussed here, it is likely that the neuronal and immune cell losses in these NID syndromes are modulated in part by cells utilizing their p53-mediated suppressor pathways.

The neuronal death in these NID syndromes, however, is unique in that the highly metabolically active neurons are the only terminally differentiated cells that may die in a few minutes when deprived of food or a reservoir for their electron flow. Since this is too short a time for transcription to be an effective mediator, neurons have had to develop other means (especially antioxidants) to protect themselves against their high rates of oxidation-prone metabolic activity. Active neurons also appear to depend on their more numerous astrocytic support cells to maintain their synapses at an optimal ionic, redox, and neurotransmitter status.[79] With sudden breaks in this astrocyte-mediated support, neurons likely die an accelerated oxidative-type death marked by rapid oxidative damage to their membrane lipids with cell swelling and lysis. However, if the astrocyte dysfunction is mild or chronic, neurons may have sufficient energy and time to activate their defensive p53-mediated transcription pathways and thus control their oxidative stresses and ceramide-mediated protease pathways to apoptotic degeneration and survive. With more severe or persistent injuries, neuronal apoptotic pathways become activated, however. Thus, in the chronic NID syndromes discussed here, senescent, necrotic, and apoptotic types of neuronal cell death often occur simultaneously.

The fate of astrocytes in these NID syndromes is primarily to differentiate. Upon infection with retroviruses or upon cellular overexpression of mitogenic cytokines (see chapters 4 and 6); some astrocytes may quickly die an early apoptotic death, whereas others may survive and differentiate into nonsupportive, slow-growing macrophagelike cells that at a later date may undergo an oxidative senescent-type (aging) death.[59] A role for p53 in the fate of these damaged astrocytes has not yet been identified.

The role of p53 in lymphocytes in these NID syndromes is also unclear. The immature lymphoid cells in thymus will rapidly die an early apoptotic death when mitogenically overstimulated by either mitogenic retroviruses or by mitogens.[11] This apoptotic death in lymphocytes is accelerated in mice lacking an optimum p53 response, as in ATm-deficient mice.[78] In p53-knockout mice, with complete absence of p53, an overgrowth of immature thymic cells with formation of multiple tumors is the result.[80] However, chronic mitogenic overstimulation, as by retroviruses, of mature lymphoid cells with an intact p53 responsiveness, leads either to a slow, redox-sensitive, senescent type of cell death with cell swelling and oxidative damage or to rapid apoptotic death.[11] Thus, the fate of lymphoid cells in NIDs depends to some extent on their ability to respond transcriptionally via their p53-controlled checkpoints to mitogenic stress, and finally on their ability to maintain a proper balance between their mitogenic pathways and their various cell death pathways.

In the NID syndromes discussed here, the uncontrolled mitogenic stimulation generated by viral products or defective genes or by inflammatory cytokines usually leads to a rapid death in peripheral immune cells but a proliferative/differentiative response by the central defense cells, i.e., astrocytes. In the face of massive amounts of mitogenic or inflammatory cytokines in CNS, the excessive stimulation to the defensive/support cells in the CNS leads to disruption of the blood-brain barrier and influx of cytotoxic lymphoid and macrophage cells into CNS and resultant killing of neurons (for details see chapters 6 and 7).

Conclusion—Pathogenesis and Potential Therapies

It has become increasingly apparent that the immune/support system both central and peripheral is etiologically involved in many of the NID syndromes outlined in chapter 2. In the CNS, either deficiencies or overactivities of the immune/support system can lead to local losses of neurons with or without influx of peripheral inflammatory cells.

Though the pathogenesis of most NID syndromes remains unclear, it appears that mutated genes that control cell fates and retroviral infections are the major etiologic factors. In either case an energy-, or redox-, or growth-factor deficiency is created that triggers either proliferative or cell death pathways. The type of cell death in CNS with NID is often a mixture of apoptosis, necrosis, and oxidative (senescent) cell death, depending on the severity and persistence of the injury. With mild and chronic damage caused by ischemia, growth factor deprivation, or energy starvation, various inhibitors of the proapoptotic proteases (caspases), including the Bcl-2 family,[29] the cowpox virus serpin,[81] or negative mutants of interleukin-1β-converting enzyme (ICE),[82,83] may be induced and temporarily delay the neuronal death. Neuronal death can also be delayed, by protease inhibitors that modulate protease activities responsible for protein processing in ER,[47,49] or by antiproteases that block activation of sphingomyelinase at the plasma membrane[2] or by deleting the caspase-sensitive protein gelsolin responsible for depolymerization of actin.[61] However, when faced with decreases in their energy supply or in their internal ionic metabolism, especially in Ca^{2+} or K^+ metabolism, or with accumulation of partially digested fibrillary peptides and/or GD3 gangliosides in ER, neurons rapidly disassemble their cytoskeleton, develop intracellular vacuoles and membrane blebs and fragment their nuclei. Because local astroglial phagocytes in CNS are scarce and unable to remove all of the soluble fibrillary proteins produced by the injuries, the accumulation of these inflammatory proteins as plaques in the CNS is inevitable.

The loss of neuronal support as a result of impaired support functions of astrocytes in NIDs may also account for much of the neuronal death seen in these syndromes. The astrocyte dysfunction in these gene or virus-induced NID syndromes may involve inhibition of the inducible astrocytic glucose transporters[16,17] that are required to supply energy and redox homeostasis both for themselves and for the neurons.

The wasting seen in these NID syndromes is likely secondary to malabsorption caused by faulty postnatal development of intestinal epithelia and/or the intestinal defense system as seen in mice deficient in cathepsin D.[47]

Although successful treatment of these various defective gene or virus-induced NID syndromes remains problematic, potential therapies focused at the central defense system, especially the culpable astrocytes in the CNS, should be productive. Since the neuronal death in these NID syndromes is thought to be caused by lack of support by their astrocytes and because astroctyes which form part of the blood-brain barrier are responsive to most extracellular immune/inflammatory signals, including most cytokines, permeant antiproteases, antioxidants, and nutrients, it may become possible to use these external modulators to maintain neuronal support functions of astrocytes under in vivo conditions. The many murine model systems of various NID syndromes now available should make the ongoing searches for such effective modulators of the astrocyte defense/support functions feasible.

Acknowledgments

We thank C. McKinley and M. Lynn for their assistance in preparing the manuscript. We also thank Richard Wilcox and Jude Richard for critical reading of the manuscript. This work is supported by Public Health Service grants CA45124 from the National Cancer Institute (NCI), AI28283 from the National Institute of Allergy and Infectious Diseases, MH57181 from the National Institute of Mental Health, Core Grant 16672 from the NCI, and a grant from the AT foundation, Austin, TX.

References

1. Thannickal VJ, Fanburg BL. Activation of an H_2O_2-generating NADH oxidase in human lung fibroblasts by transforming growth factor β1. J Biol Chem 1995; 270:30334-30338.

2. Mansat V, Bettaieb A, Levade T et al. Serine protease inhibitors block neutral sphingomyelinase activation, ceramide generation, and apoptosis triggered by daunorubicin. FASEB J 1997; 11:695-702.

3. Wright SC, Zheng H, Zhong J. Tumor cell resistance to apoptosis due to a defect in the activation of spingomyelinase and the 24 kDa apoptotic protease (AP24). FASEB J 1996; 10:325-332.

4. Los M, Van de Craen M, Penning LC et al. Requirement of an ICE/CED-3 protease for Fas/APO-1-mediated apoptosis. Nature 1995; 375:81-83.

5. Yu SP, Yeh C-H, Sensi SL et al. Mediation of neuronal apoptosis by enhancement of outward potassium current. Science 1997; 278:114-117.

6. Yee NS, Paek I, Besmer P. Role of *kit*-ligand in proliferation and suppression of apoptosis in mast cells: Basis for radiosensitivity of *white spotting* and *steel* mutant mice. J Exp Med 1994; 179:1777-1787.

7. Argiles JM, Lopez-Soriano J, Busquets S et al. Journey from cachexia to obesity by TNF. FASEB J 1997; 11:743-751.

8. Nakagawara A, Nakamura Y, Ikeda Hl et al. High levels of expression and nuclear localization of interleukin-1 β converting enzyme (ICE) and CPP32 in favorable human neuroblastomas. Cancer Res 1997; 57:4578-4584.

9. He J, DeCastro CM, Vandenbark GR et al. Astrocyte apoptosis induced by HIV-1 transactivation of the c-*kit* protooncogene. Proc Natl Acad Sci 1997; 94:3954-3959.

10. Haraguchi S, Good RA, James-Yarish M et al. Induction of intracellular cAMP by a synthetic retroviral envelope peptide: A possible mechanism of immunopathogenesis in retroviral infections. Proc Natl Acad Sci USA 1995; 92:5568-5571.

11. Lynn WS, Wong PKY. Neuroimmunopathogenesis of ts1 MoMuLV viral infection. NeuroImmunoModulation 1998; in press.

12. Saha K, Yuen PH, Wong PKY. Murine retrovirus-induced destruction of $CD4^+$ T cells and thymocytes are mediated through activation-induced death by apoptosis. J Virol 1994; 68:2735-2740.

13. Walev I, Reske K, Palmer M et al. Potassium-inhibited processing of IL-1β in human monocytes. EMBO J 1995; 14:1607-1614.

14. Hannun YA. Functions of ceramide in coordinating cellular responses to stress. Science 1996; 274:1855-1861.

15. Piper RC, James DE, Slot JW et al. GLUT-4 phosphorylation and inhibition of glucose transport by dibutryl cAMP. J Biol Chem 1993; 268:16557-16563.

16. Berridge MV, Tan AS, McCoy KD et al. CD95 (Fas/Apo-1)-induced apoptosis results in loss of glucose transporter function. J Immunol 1996; 156:4092-4099.

17. Kan O, Baldwin SA, Whetton AD. Apoptosis is regulated by the rate of glucose transport in an interleukin 3 dependent cell line. J Exp Med 1994; 180:917-923.

18. Marthinuss J, Andrade-Gordon P, Seiberg M. A secreted serine protease can induce apoptosis in Pam212 keratinocytes. Cell Growth & Differ 1995; 6:807-816.

19. Ryser HJ-P, Levy EM, Mandel R et al. Inhibition of human immunodeficiency virus infection by agents that interfere with thiol-disulfide interchange upon virus-receptor interaction. Proc Natl Acad Sci USA 1994; 91:4559-4563.

20. Lawrence DA, Song R, Weber P. Surface thiols of human lymphocytes and their changes after in vitro and in vivo activation. J Leukoc Biol 1996; 60:611-618.

21. Buttke TM, Sandstrom PA. Oxidative stress as a mediator of apoptosis. Proc Natl Acad Sci USA 1994; 90:10061-10065.

22. Kopczynski CC, Davis GW, Goodman CS. A neural tetraspanin, encoded by late bloomer, that facilitates synapse formation. Science 1996; 271:1867-1870.

23. Wakade AR, Przywara DA, Palmer KC et al. Deoxynucleoside induces neuronal apoptosis independent of neurotrophic factors. J Biol Chem 1995; 270:17986-17992.

24. Liu X, Kim CN, Yang J et al. Induction of apoptotic program in cell-free extracts: Requirement for dATP and cytochrome c. Cell 1996; 86:147-157.

25. Yang J, Liu X, Bhalla K et al. Prevention of apoptosis by Bcl-2: Release of cytochrome c from mitochondria blocked. Science 1997; 275:1129-1132.

26. Kluck RM, Bossy-Wetzel E, Green DR et al. The release of cytochrome c from mitochondria: A primary site for Bcl-2 regulation of apoptosis. Science 1997; 275:1132-1137.

27. Eguchi T, Shimizu S, Tsujimoto Y. Intracellular ATP levels determine cell death fate by apoptosis or necrosis. Cancer Res 1997; 57:1835-1840.

28. Brand KA, Hermfisse U. Aerobic glycolysis by proliferating cells: A protective strategy against reactive oxygen species. FASEB J 1997; 11:388-395.

29. Shimizu S, Eguchi Y, Kamiike W et al. Bcl-2 expression prevents activation of the ICE protease cascade. Oncogene 1996; 12:2251-2257.

30. Sundaresan M, Yu Z-X, Ferrans VJ et al. Requirement for generation of H_2O_2 for platelet-derived growth factor signal transduction. Science 1995; 270:296-299.

31. Antonsson B, Conti F, Ciavatta AM et al. Inhibition of Bax channel-forming activity by Bcl-2. Science 1997; 277:370-372.

32. Venable ME, Lee JY, Smyth MJ et al. Role of ceramide in cellular senescence. J Biol Chem 1995; 270:30701-30708.

33. Cifone MG, Roncaioli P, DeMaria R et al. Multiple pathways originate at the Fas/APO-1 (CD95) receptor: Sequential involvement of phosphatidylcholine-specific phospholipase C and acidic sphingomyelinase in the propagation of the apoptotic signal. EMBO J 1995; 14:5859-5868.

34. Schulze-Osthoff K. The Fas-APO-1 receptor and its deadly ligand. Trends Cell Biol 1994; 4:421-429.

35. Choe WK, Stoica G, Lynn WS et al. Neurodegeneration induced by MoMuLV-*ts*1 and increased expression of TNFα and Fas in the central nervous system. Brain Res 1998; 779:1-8.

36. Wesselingh SL, Tyor WR, Griffin DE. Cytokines and HIV-associated dementia. In: Ransohoff RM and Benveniste EN eds. Cytokines and the CNS. Boca Raton: CRC Press, 1996:287-307.

37. Ezzell C. Protein folding and the early secretory pathway: Researchers begin to understand the cellular assembly line. J NIH Res 1997; 9:42-47.

38. Loo TW, Clarke DM. Correction of defective protein kinesis of human P-glycoprotein mutants by substrates and modulators. J Biol Chem 1997; 272:709-712.

39. Diedrich JF, Bendheim PE, Kim YS et al. Scrapie-associated prion protein accumulates in astrocytes during scrapie infection. Proc Natl Acad Sci USA 1991; 88:375-379.

40. Selkoe DJ. Alzheimer's disease: Genotypes, phenotype, and treatments. Science 1997; 275:630-632.

41. Gonzales-Scarano F, Nathanson N, Wong PKY. Retroviruses and the nervous system, In: Levy JA ed. The Retroviridae, vol. 4. New York: Plenum Press, 1995:409-490.

42. Pahl HL, Sester M, Burgert H-G et al. Activation of transcription factor NF-κB by the adenovirus E3/19K protein requires its ER retention. J Cell Biol 1996; 132:511-522.

43. Walker KW, Gilbert HF. Scanning and escape during protein-disulfide isomerase-assisted protein folding. J Biol Chem 1997; 272:8845-8848.

44. De Maria R, Lenti L, Malisan F et al. Requirement for GD3 ganglioside in CD95- and ceramide-induced apoptosis. Science 1997; 277:1652-1656.

45. Enari M, Hug H, Nagata S. Involvement of an ICE-like protease in Fas-mediated apoptosis. Nature 1995; 375:78-81.

46. Lazebnik YA, Kaufmann SH, Desnoyers S et al. Cleavage of poly(ADP-ribose) polymerase by a proteinase with properties like ICE. Nature 1994; 371:346-347.

47. Saftig P, Hetman M, Schmahl W et al. Mice deficient for the lysosomal proteinase cathepsin D exhibit progressive atrophy of the intestinal mucosa and profound destruction of lymphoid cells. EMBO J 1995; 14:3599-3608.

48. Shikova E, Lin Y-C, Saha K et al. Astrocyte-specific defective gPr80^{env} processing correlates with cytopathogenicity induced by *ts1*, a mutant of Moloney murine leukemia virus. J Virol 1993; 67:1137-1147.

49. Deiss LP, Galinka H, Berissi H et al. Cathepsin D protease mediates programmed cell death induced by interferon-γ, Fas/APO-1 and TNF-α. EMBO J 1996; 15:3861-3870.

50. Kamps CA, Lin Y-C, Wong PKY. Oligomerization of the envelope protein of Moloney murine leukemia virus-TB and of *ts1*, a neurovirulent temperature-sensitive mutant of MoMuLV-TB. Virology 1991; 184:687-694.

51. Yu Y, Kamps CA, Yuen PH et al. Construction and characterization of expression systems for the *env* gene of *ts1*, a mutant of Moloney murine leukemia virus-TB. Virus Res 1991; 19:83-92.

52. Szurek PF, Yuen PH, Ball JK et al. A Val-25-to-Ile substitution in the envelope precursor polyprotein, gPr80^{env}, is responsible for the temperature sensitivity, inefficient processing of gPr80^{env}, and neurovirulence of *ts1*, a mutant of Moloney murine leukemia virus TB. J Virol 1990; 64:467-475.

53. Westphal CH, Schmaltz C, Rowan S et al. Genetic interactions between atm and p53 influence cellular proliferation and irradiation-induced cell cycle checkpoints. Cancer Res 1997; 57:1664-1667.

54. Baskaran R, Wood LD, Whitaker LL et al. Ataxia telangiectasia mutant protein activates c-Abl tyrosine kinase in response to ionizing radiation. Nature 1997; 387:516-519.

55. Xu Y, Baltimore D. Dual roles of ATM in the cellular response to radiation and in cell growth control. Genes Dev 1996; 10:2401-2410.

56. Kim H-RC, Upadhyay S, Li G et al. Platelet-derived growth factor induces apoptosis in growth-arrested murine fibroblasts. Proc Natl Acad Sci USA 1995; 92:9500-9504.

57. Meyn MS. Ataxia-telangiectasia and cellular responses to DNA damage. Cancer Res 1995; 55:5991-6001.

58. Xu Y, Ashley T, Brainerd EE et al. Targeted disruption of ATM leads to growth retardation, chromosomal fragmentation during meiosis, immune defects, and thymic lymphoma. Genes Dev 1996; 10:2411-2422.

59. Sugrue MM, Shin DY, Lee SW et al. Wild-type p53 triggers a rapid senescence program in human tumor cells lacking functional p53. Proc Natl Acad Sci USA 1997; 94:9648-9653.

60. Chen X, Ko LJ, Jayaraman L et al. p53 levels, functional domains, and DNA damage determine the extent of the apoptotic response of tumor cells. Genes Dev 1996; 10:2438-2451.

61. Kothakota S, Azuma T, Reinhard C et al. Caspase-3-generated fragment of gelsolin: Effector of morphological change in apoptosis. Science 1997; 278:294-298.

62. Martin SJ, Finucane DM, Amarante-Mendes GP et al. Phosphatidylserine externalization during CD95-induced apoptosis of cells and cytoplasts requires ICE/CDE-3 protease activity. J Biol Chem 1996; 271:28753-28756.

63. Levine A, Tenhaken R, Dixon R et al. H_2O_2 from the oxidative burst orchestrates the plant hypersensitive disease resistance response. Cell 1994; 79:583-593.

64. Beal MF. Mitochondrial dysfunction and oxidative damage in neurodegenerative diseases. Austin, TX: RG Landes Company 1995.

65. Floyd RA, Carney JM. Nitrone Radical Traps Protect In Experimental Neurodegenerative Diseases, Free Radical Biology & Aging Research. Academic Press Limited, Oklahoma City, OK 1996:70-83

66. Lipton SA, Choi Y-B, Pan Z-H et al. A redox-based mechanism for the neutroprotective and neurodestructive effects of nitric oxide and related nitroso-compounds. Nature 1993; 364:626-631.

67. Talley AK, Dewhurst S, Perry SW et al. Tumor necrosis factor alpha-induced apoptosis in human neuronal cells: Protection by the antioxidant *N*-acetylcysteine and the genes *bcl-2* and *crmA*. Mol Cell Biol 1995; 15:2359-2366.

68. Fadok VA, Voelker DR, Campbell PA et al. Exposure of phosphatidylserine on the surface of apoptotic lymphocytes triggers specific recognition and removal by macrophages. J Immunol 1992; 148:2207-2216.

69. Polymeropoulos MH, Lavedan C, Leroy E et al. Mutation in the α-synuclein gene identified in families with Parkinson's disease. Science 1997; 276:2045-2047.

70. Jung M, Zhang Y, Lee S et al. Correction of a radiation sensitivity in ataxia telangiectasia cells by a truncated IκB-α. Science 1995; 268:1619-1621.

71. Jung M, Kondasatyev A, Sung AL et al. ATm gene phosphorylates IκBα. Cancer Res 1997; 57:24-27.

72. Gottlieb MT, Oren M. p53 in growth control and neoplasia. Biochem Biophys Acta 1996; 1287:77-102.

73. Haraguchi S, Good RA, Cianciolo GJ et al. Immunosuppressive retroviral peptides: Immunopathological implications for immunosuppressive influences of retroviral infections. J Leukoc Biol 1997; 61:654-666.

74. Levy JA. Concepts in HIV neuropathogenesis. In: Ncu HC, Levy JA, Weiss R eds. Focus on HIV. London, England: Churhill Livingstone, 1993:51-67.

75. Serrano M, Lin AW, McCurrach ME et al. Oncogenic *ras* provokes premature cell senescence associated with accumulation of p53 and p16^{INK4a}. Cell 1997; 88:593-602.

76. Ko LJ, Prives C. p53: Puzzle and paradigm. Genes Dev 1996; 10:1054-1072.

77. Hollstein M, Shomer B, Greenblatt M et al. Somatic point mutations in the p53 gene of human tumors and cell lines: Updated compilation. Nucleic Acids Res 1996; 24:141-146.

78. Barlow C, Hirotsune S, Paylor R et al. Atm-deficient mice: A paradigm of ataxia telangiectasia. Cell 1996; 86:159-171.

79. Tsacopoulos M, Magestretti PJ. Metabolic coupling between glia and neurons. J Neurosci 1996; 16:877-885.

80. Hursting SD, Perkins SN, Brown CC et al. Calorie restriction induces a p53-independent delay of spontaneous carcinogenesis in p53-deficient and wild-type mice. Cancer Res 1997; 57:2843-2846.

81. Gagliardini V, Fernandez P-A, Lee RKK et al. Prevention of vertebrate neuronal death by the crmA gene. Science 1994; 263:826-828.

82. Friedlander RM, Brown RH, Gagliardini V et al. Inhibition of ICE slows ALS in mice. Nature 1997; 388:31.

83. Friedlander RM, Gagliardini V, Hara H et al. Expression of a dominant negative mutant of interleukin 1β converting enzyme in transgenic mice prevents neuronal cell death induced by trophic factor withdrawal and ischemic brain injury. J Exp Med 1997; 185:933-940.

*ts*1 MoMuLV: A Murine Model of Neuroimmunodegeneration

Paul K.Y. Wong, William S. Lynn, Y.C. Lin, Wonkyu Choe and P.H. Yuen

Introduction

Replication-competent retrovirus infection results either in accelerated cell proliferation leading to tumorigenesis or to degenerative cell death, particularly in the immune and nervous systems. The outcome depends on the virus, the cell type affected, and the stage of cell differentiation. The Moloney murine leukemia virus (MoMuLV) family is a good example of a retrovirus that can produce either outcome in the infected host. Although wild-type (WT) MoMuLV primarily causes T cell lymphoma, several temperature-sensitive (*ts*) mutants of MoMuLV have been shown to cause early and fatal degenerative disorders of the immune and nervous systems when administered to neonatal mice.[1] One of these *ts* mutants, designated *ts*1 MoMuLV, has been studied intensively. *ts*1 MoMuLV, isolated in 1973,[2] was the first neuroimmunopathogenic retrovirus to be isolated in vitro from a nonneuroimmunovirulent MuLV.[3] *ts*1 MoMuLV and its variants, however, are not the only murine retroviruses that induce neurological disorders. A large number of murine retroviruses, including CasBrE MuLV, an isolate of MuLV of wild mouse origin, are capable of inducing neurological disorders.[4] Another murine retrovirus, LP-BM5, the virus which induces severe T and B cell deficiencies (murine AIDS), also induces a mild form of neurological disorder (see chapter 5). *ts*1 MoMuLV, however, is unique among murine retroviruses in that it mediates both severe T cell and neuronal cell losses during postnatal development, resulting in

Neuroimmunodegeneration, edited by Paul K.Y. Wong and William S. Lynn.
© 1998 Springer-Verlag and R.G. Landes Company.

drastic disorders involving both the immune and nervous systems. The cause of this neuroimmunodegeneration (NID) syndrome has been attributed to two point mutations in the *env* gene. These two mutations result in a Val→Ile substitution at position 25 and an Arg→Lys substitution at position 430 of the envelope polypeptide gPr80^{env}. The Val 25→Ile substitution renders the newly synthesized envelope protein unable to fold properly at the nonpermissive temperature of 39°C. As a consequence, the misfolded precursor envelope protein gPr80^{env} oligomerizes in the endoplasmic reticulum (ER) but fails to be transported out and thus accumulates there.[5-7] Interestingly, in certain cell types such as fibroblasts and endothelial cells, this inefficient transport and processing of gPr80^{env} in the ER is seen only at the nonpermissive temperature, whereas in other cell types, such as T cells and astrocytes, this inefficient transport and processing of gPr80^{env} occurs at both the permissive and nonpermissive temperatures.[8,9] Using confocal microscopy to study the distribution of envelope proteins in several infected cell types, we have observed that, at the permissive temperature of 34°C, the accumulation of gPr80^{env} in astrocytes[8] and T cells is perinuclear (unpublished data). This apparent distribution in the ER/Golgi apparatus is not seen at 34°C in infected endothelial cells.[8] This finding is consistent with the notion that transport of the precursor envelope proteins is inefficient only in cell types such as astrocytes and T cells, which are killed by this virus, but not in endothelial cells which are not killed by *ts*1 MoMuLV infection. Our hypothesis for how *ts*1 envelope causes NID is that accumulation of the unprocessed precursor envelope protein, alone or in complexes with other cellular proteins, clogs the ER. This in turn either blocks transport to the cell surface of essential cellular proteins required for cell survival or induces excessive liberation from the ER of apoptotic signals such as Ca^{2+} or the ganglioside GP3, which activates the caspase or the ceramide-mediated pathway to apoptotic cell death. Thus, this specific virus-cell interaction illustrates the point that the outcome of the viral infection depends not only on the virus (in this case a retrovirus with a mutation in the *env* gene) but also on the cell type involved. The results of the Arg430→Lys substitution in the *ts*1 MoMuLV envelope[5] is not as well defined, but it appears to alter the binding efficiency of the envelope protein to specific cell types. It also may enhance the ability of the virus to act as a mitogen and activate both cell proliferation and viral production.

That the *ts1 env* gene is responsible for the neuropathogenicity of *ts1* was confirmed recently in our studies of transgenic mice, which showed that expression of the *ts1 env* gene in mice alone can cause neurodegeneration, albeit milder than that seen in *ts1*-infected mice.[10]

Clinical Features

Infection of newborn FVB/N mice, the most susceptible mouse strain,[11] with *ts1* MoMuLV results in a progressive NID syndrome in 100% of the mice. The mice invariably die 35 to 45 days postinfection (dpi). Clinical signs begin around 21 dpi with slight body tremor. By 20 to 25 dpi, mice become scruffy and develop ataxia, apathy, severe wasting, and paraparesis that progresses to hind limb paralysis and death. These mice also develop severe thymic atrophy with loss of 80 to 90% of their thymocytes.[12] The cell count in the spleen also decreases as the disease progresses. The splenic atrophy, however, is not as severe as thymic atrophy. A few of the infected, wasted mice also develop severe diarrhea, the cause of which has not been fully evaluated but may be associated with secondary infection due to immunosuppression. Because these wasted mice eat and drink normally as well as grow normally until after weaning, the wasting of these neonatal mice is likely the result of malabsorption of calories, which develops only during the postweaning period. Similar clinical features and symptoms are observed in other susceptible strains of mice infected neonatally with *ts1* MoMuLV, including Balb/c, CFW/D, and C3H.[3] However, the time of onset of wasting, paralysis, and death in these strains of mice is more variable. Some strains, e.g., C57/BL, are completely resistant to *ts1* MoMuLV-induced NID.[3]

Potential Pathogenic Mechanisms

Causes of Cell Losses in the Immune System

With intraperitoneal infection, *ts1* replicates initially in the thymus and spleen, particularly in thymocytes and $CD4^+$ cells, before spreading to the CNS. In the thymus, several cell types are infected, including dendritic cells, endothelial cells, and thymocytes, but the cytopathic effect is observed mainly in the immature thymocytes.[12] During the early phase of *ts1* infection, there is an increase in thymocytic mitotic figures, followed by progressive thymocytic cell death, particularly in the cortex, and severe thymic atrophy by 20-30 dpi (Fig. 4.1). Depletion of lymphocytes in the infected mice is

Fig. 4.1. (A) numbers of mitotic cortical thymocytes in *ts1* MoMuLV-infected (●) and control (□) FVB/N mice at various time, post infection. Each bar represent mean ± SE of counts from six 40x fields in sections from two mice. (B) numbers of pyknotic (apoptotic) thymocytes in *ts1* MoMuLV-infected (●) and control (□) FVB/N mice. Each bar represents the mean ± SE of counts from six 40x fields in sections from two mice. (C) thymic weight from an average of 10 *ts1* MoMuLV-infected (●) and 10 control (□) FVB/N mice. Bar represents the mean ± SE.

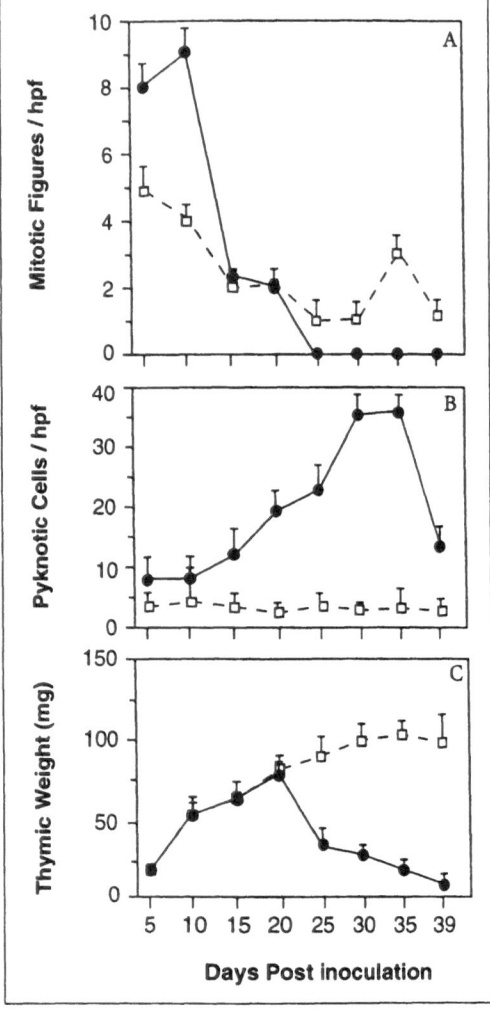

also seen in the peripheral blood.[13] The killing of thymocytes by *ts1* appears to be a direct result of viral infection of the thymocytes.[14] The type of cell death is apoptotic, as indicated by fragmentation of DNA, chromatin condensation, extensive blebbing and vesiculation at the plasma membrane and by the observation that death is blocked by protein synthesis inhibitors.[15] Thymocytes freshly isolated from *ts1*-infected mice at 20-25 dpi, mainly double-positive ($CD4^+CD8^+$) immature cells, also appear to be mitogenically activated in vitro as they are in vivo.[12,15] However, the mitogenically activated cells rapidly die when cultured in serum-deficient media with fragmented

DNA, condensation of chromatin, and vesiculation. The more mature splenocytes obtained from these infected mice are also initially mitogenically activated. However, they rapidly shrink, excrete vesicles containing DNA, and rapidly die under serum-deficient conditions.[16] These findings indicate that in vivo infection by ts1 MoMuLV results initially in mitogenic overactivation of both mature and immature lymphocytes, which rapidly leads to apoptotic cell death. In other studies, this accelerated vesicular cell death of ts1-infected splenocytes in low serum could be prevented by the early addition of appropriate growth, survival, and redox factors.[16,17] Th1 cytokines, primarily IL-2, IL-12, and IFNγ, plus redox support, i.e., N-acetyl cysteine, melatonin, or dimethyl sulfoxide, were shown to prolong cell survival and prevent activation of the splenic cell death pathways. Thymocyte survival, however, required Th2 cytokines, i.e., IL-4 and IL-6. In contrast, antioxidants were not protective in thymocytes. Rather, the redox agents accelerated the cell death pathways in the infected thymocytes. These findings suggest that ts1 infection causes the accumulation in vivo of mitogenically active abnormal cells that more rapidly activate their apoptotic pathways when further mitogenically stimulated in the absence of serum. Because in vitro application of the appropriate mix of survival, growth, and redox factors was shown to prevent cell death in these apoptosis-prone cells, it appears that these cells fail to survive in vivo or in vitro because, under mitogenic stress, they are unable to produce adequate amounts of their own specific survival and growth factors. These findings raise the question of how ts1 infection interferes with production of essential survival factors by these infected lymphoid cells. Our previous data, as outlined above, showing that the mutated envelope protein of ts1 MoMuLV is inefficiently transported and accumulates in the ER, suggests an answer. Because most cytokine growth factors must be processed into activated forms in the ER before exit to the plasma membrane, the accumulation of excessive amounts of foreign, insoluble protein such as gPr80^{env} in the ER may result in competitive inhibition of transport and processing of these essential factors required for cell survival.

To further investigate the specific events leading to T cell depletion by ts1, we have established a long term primary culture system using dissociated thymocytes.[9] These thymocytes retain their double-positive CD4$^+$CD8$^+$ immature phenotype in culture and are susceptible to infection by ts1 and by WT MoMuLV. As in our ex vivo

studies, these in vitro *ts1*-infected thymocytes also proliferate initially at an accelerated rate but subsequently die much faster than
control or WT-infected thymocytes. Fragmented DNA appears in
these *ts1*-infected cells as early as 8 hours after infection but not in
WT-infected cells or uninfected control cells until later (data not
shown). By 72 hours postinfection, WT-infected cells contain some
fragmented DNA, but much less than that seen in *ts1*-infected cells
at the same time point. Thus, our in vitro studies mimic those seen
in *ts1*-infected T cells in vivo.[15] To examine whether the inefficient
processing of gPr80[env] that is seen in the thymus of *ts1*-infected mice[15]
also occurs in these primary thymocytes, we compared the kinetics
of processing of gPr80[env] in *ts1*-infected and WT-infected thymocytes.
However, the relatively low level of virus expression in primary thymocytes made the detection of viral envelope protein difficult. As
shown in Figure 4.2A (lane 1), a 15 min pulse, which is normally used
to obtain a readily visible band in infected primary cultures of endothelial cells (Fig. 4.2C, lane 1), or astrocytes[8] yielded no visible
band in primary culture of thymocytes. But with a longer pulse of
up to 2 hours, a visible band was detected (Fig. 4.2A, lane 2). Therefore, to study the kinetics of gPr80[env] processing in thymocytes, a
longer pulse of 4 hours was used. As shown in Figure. 4.2B, a gPr80[env]
band was observed for WT-infected thymocytes after a 4 hour pulse
(0 hour chase, lane 1); however, there was a significant reduction of
the gPr80[env] band after 2 hour of chase (Fig. 4.2B, lane 2) and by the
4 hour chase, the band completely disappeared (data not shown). In
ts1-infected thymocytes, a band more intense than that for the
WT-infected cells was observed after a 4 hour pulse (Fig. 4.2B 0 hour
chase, lane 3), consistent with the higher viral titer of *ts1*-infected
thymocytes compared to WT-infected thymocytes.[18] Although there
was a decline in the intensity of the *ts1* gPr80[env] band from 0 hour to
2 hour of chase, there was no decline in the intensity of the band
from 2 hour to 4 hour of chase, and a significant portion of the
gPr80[env] remained inside the cells even after a 4 hour chase. These
results suggest that, although some gPr80[env] in *ts1*-infected cells could
be transported out and processed, a significant portion of gPr80[env]
remained inside the cells even after 4 hours of chase. They also suggest that the clogging of primary thymocytes in the ER by the accumulating gPr80[env] may interfere with the processing and export
of essential growth survival factors such as IL-4 and IL-6 that are
required to suppress the apoptotic cell-death pathways in these
mitogenically activated thymocytes.

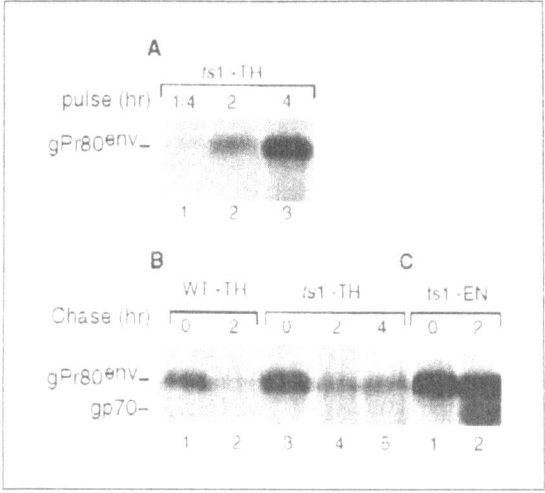

Fig. 4.2. Pulse-chase analysis of primary thymocytes infected with *ts1* or WT viruses. Analyses were performed as specified below: (A) *ts1*-infected thymocytes were pulsed for different time periods as indicated; (B) *ts1*- or WT-infected (2-hour chase); lane 3, *ts1*-infected (2-hour chase); lane 5, *ts1*-infected (4-hour chase); and (C) *ts1*-infected primary endothelial cell were pulsed for 15 minutes and chased for 0 hours (lane 1) and 2 hours (lane 2). Abbreviations: TH = thymus; EN= endothelial cells.

Causes of Cell Losses in the CNS

The most prominent pathologic features of *ts1* infection in the CNS are neuronal loss, demyelination, gliosis, and spongiform lesions with no detectable inflammatory cell infiltration.[19] Spongiform degeneration is found mainly in the spinal cord (SC) and brain stem (BS); to a lesser degree in the cerebellum, thalamus, and hypothalamus; and rarely in the cerebral cortex (CC). In the early stage, intracellular vacuolation generally begins in a perivascular and perineuronal location. As the disease progresses, these vacuoles increase in size and number.[4] Apoptotic neurons containing condensed and dispersed nuclei are occasionally found in the vicinity of the large vacuoles.[4] Apoptotic changes are also seen in some astrocytes. Other astrocytes show mitochondrial swelling and intracellular vacuolization.[19] The virus titer as well as the intensity of the pathology in the affected region increases as the disease progresses. In late-stage disease, neuron counts in the affected area of *ts1*-infected mice are significantly lower than counts in similar areas in uninfected controls. In the affected areas, although neurons are most severely damaged, viral antigens are primarily found in endothelial and glial cells but not in neurons.[19] These findings suggest that the neuronal death is caused by either impairment of those helper functions of infected glial cells that are essential for neuronal survival or by the inappropriate or excessive production of neurotoxic factors by the infected glial cells, e.g., TNF, free radicals, quinolinic acids, arachidonic acid,[19a] and *ts*1 MoMuLV envelope peptides.[20-22] To evaluate the possibility that the overexpression of neurotoxic proinflammatory cytokines or proteins is responsible for the neuronal losses, measurements of cytokines as well as apoptotic and antiapoptotic factors in the degenerating area of the CNS have been carried out.[23] Our results showed that TNFα and IL-1 were expressed at levels much higher than uninfected controls in *ts*1-infected BS, where the spongiform lesions occurred (Fig. 4.3). Expression of these cytokines in the CC, where spongiform degeneration was rarely seen, was minimal (Fig. 4.3). TNFα expression in BS increased as disease progressed, indicating that in *ts*1-infected mice, TNFα expression correlates well with disease location and progression. Although the increased expression of IL-1 in the BS also correlated with disease progression, the amount of mRNA of IL-1 was much less than that of TNFα (data not shown). Expression of TGFβ, Bax, and Bcl-2 was consistently high at all disease stages but did not differ in BS and CC between *ts*1-infected and healthy control mice. IL-6 expression was weak in both diseased and control CNS. No expression of IL-2, IL-4, or IFNγ was detected at any stage in the CNS of these neonatal infected mice. The absence of these cytokines, however, is not surprising since *ts*1-infected CNS lacks inflammatory lymphoid or macrophage infiltrates.[19,24] To correlate expression of the cell death-related gene product Fas with *ts*1-induced neurodegeneration, expression of this gene was determined in the BS and CC of *ts*1-infected mice. The increased expression of Fas in the BS, like the increased expression of TNF in the BS, correlated well with the intensity of clinical signs and with the localization of the CNS lesions (Fig. 4.3). Since mRNA expression of TNFα and Fas was significantly increased in the *ts*1 MoMuLV-infected BS, the expression of TNFα, Fas, and Fas ligand (FasL) protein in BS was also evaluated by immunohistochemical staining using antibodies against TNFα, Fas, and FasL.[23] All cells immunolabeled for TNFα protein were located in the areas of spongiform degeneration and gliosis. With double labeling for TNFα and glial fibrillary acidic protein (GFAP) the GFAP-positive

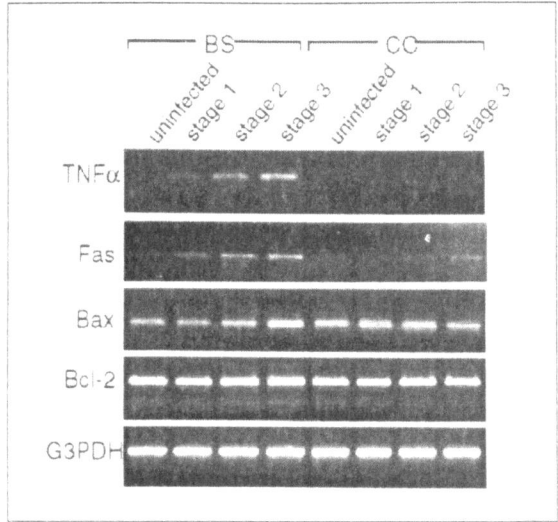

Fig. 4.3. Relative expression of mRNA for TNF, Fas, Bax, Bcl-2 in brain stem (BS) and cerebral cortex (CC) of *ts1* infected mice. To facilitate the correlation of disease progression to pathogenesis, the disease progression is classified into three stages based on the intensity of the clinical signs: Stage 1 (mild body tremor), Stage 2 (ataxia and paraparesis), and Stage 3 (hind limb paralysis).

astrocytic processes were also highly positive for TNFα. A few of the cells near the spongiform lesions stained positively for TNFα but not for GFAP, indicating that these cells could be either microglia or oligodendrocytes or unactivated or dying astrocytes. Only about 30% of the cells within the area of spongiform degeneration were TNFα positive. TNFα immunostaining was negative in the control CNS. Fas immunolabeling in the CNS of infected mice was also predominant within the BS areas of spongiform degeneration. On double labeling for Fas and GFAP, a smaller percentage (about 10-20%) of the GFAP-positive cells also stained positively for Fas. Fas staining of GFAP-negative cells was also observed. On the basis of morphology, some of these Fas-positive/GFAP-negative cells appeared to be astrocytes. A smaller number appeared to be motor neurons. Many of the Fas-positive cells also showed various degrees of degeneration characterized by cytoplasmic vacuolar changes and nuclear chromatin condensation. The cortical region of *ts1*-infected CNS and control CNS was negative for Fas immunostaining. Many cells, possibly astrocytes, at the site of spongiform lesions of the BS were also labeled with FasL, and most of the FasL-positive cells also showed various degrees of degeneration. Thus, these immunohistochemical studies demonstrated that Fas, FasL, and TNFα proteins, along with their mRNAs, are present primarily in astrocytes in the spongiform areas of the *ts1*-infected BS but not in the nonvacuolated CC.[23] To evaluate whether the *ts1*-infected BS expressed biologically active TNFα protein, brain homogenates obtained from *ts1*-infected mice were tested for TNFα bioactivity on L929 cells, a line susceptible to TNFα-mediated killing. A cytotoxic level of TNFα for L929 cells was observed in *ts1*-infected BS but not in BS of healthy control mice.[23] Moreover, treatment of *ts1*-infected BS homogenates with neutralizing anti-TNFα antibody eliminated the cytotoxicity of the *ts1*-BS homogenate on L929 cells, thus confirming the presence of the bioactive TNFα in the infected BS.[23]

Possible Mechanism(s) of Neuronal Death Mediated by *ts1* MoMuLV Infection of Astrocytes

Astrocytes are the most abundant cells in the CNS with inducible immune functions and as such are a major source of proinflammatory and neurotoxic cytokines when activated. However, when not differentiated into dendritic-like cells, astrocytes are also a major source of essential neurotrophic growth factors and essential nutrient and redox support for neurons.[22,25] But whether either the proinflammatory function or the impairment of the neuronal support function or both are responsible for neuronal losses in *ts1*-infected mice remains unclear.

Previously, using primary astrocytes, we showed that the cytopathic effects of *ts1* on astrocytes are closely correlated with accumulation of the precursor envelope protein of *ts1* in the ER.[8] As in primary astrocytes, murine astrocytes immortalized using the temperature-sensitive SV40 tsA58 T antigen showed cytopathic effects as a result of *ts1* infection.[26] These cytopathic effects included growth inhibition, exposure of inner plasma membrane phosphatidylserine, swelling and loss of cell processes with syncytium formation, excretion of vesicles containing fragmented DNA, and release of dying cells from the plastic surface of culture plates (see below). However, only about 35% of the infected cells died by 5 dpi. By 5 dpi, the remaining viable *ts1*-infected astrocytes had differentiated into slow-growing, virus-resistant, activated cells secreting high levels of TNF and IL-1 (see below). Interestingly, the transport and processing of *ts1* MoMuLV precursor envelope protein in the virus-resistant, differentiated astrocytes was much more efficient than in the infected viable cells isolated at 2-3 dpi (unpublished data). Thus, these surviving astrocytes, in response to virus infection reduced their growth rate, excreted TNF and IL-1, rapidly processed the *ts1* gPr80^{env}, and became resistant to *ts1*-induced cell death. These observations indicate that *ts1* MoMuLV infection of astrocytes in vitro can activate either apoptotic pathways or differentiation/survival pathways. These in vitro findings also parallel those seen in vivo in *ts1* MoMuLV-infected astrocytes,[19] some of which die while others either proliferate to produce gliosis or differentiate into activated macrophage-like cells secreting TNF and IL-1.[23] It appears, therefore, that the main effect of *ts1* in astrocytes is to alter rates of those various pathways that control cell fate. The final outcome for the infected astrocytes is either death or differentiation into virus-resistant and highly activated cells. Thus, *ts1* MoMuLV infection of astrocytes, which either kills astrocytes or causes them to differentiate into activated macrophage-like or dendritic cells that are unlikely to be able to maintain their normal support functions, may deprive neurons of their essential astrocyte-derived supports. With such deficiencies, the metabolically highly active neurons would then rapidly die. In addition, these differentiated astrocytes produce large amounts of potential neurotoxins as well as reduce their ability to maintain optimal concentration of excitotoxic amino acids, e.g., glutamate (see below). Thus the *ts1*-induced neuronal degeneration seen in vivo is likely due to loss of astrocytic support or the production of neurotoxins or both by the infected astrocytes.

Alteration of Signaling Pathways at the Plasma Membrane by *ts1 MoMuLV* Infection

Exposure of Phosphatidylserine at the Plasma Membrane

Since HIV envelope proteins have recently been shown to perturb plasma membrane ion transport in astrocytes,[27] it is possible that the *ts1* envelope may also perturb the plasma membrane functions of astrocytes. To evaluate this possibility, changes in plasma membrane phospholipid orientation during the course of *ts1* infection, as monitored by spectral changes in the membrane-bound fluorochrome MC540, were observed. The *ts1*-infected astrocytes were shown to bind increased amounts of MC540 (Fig. 4.4) before

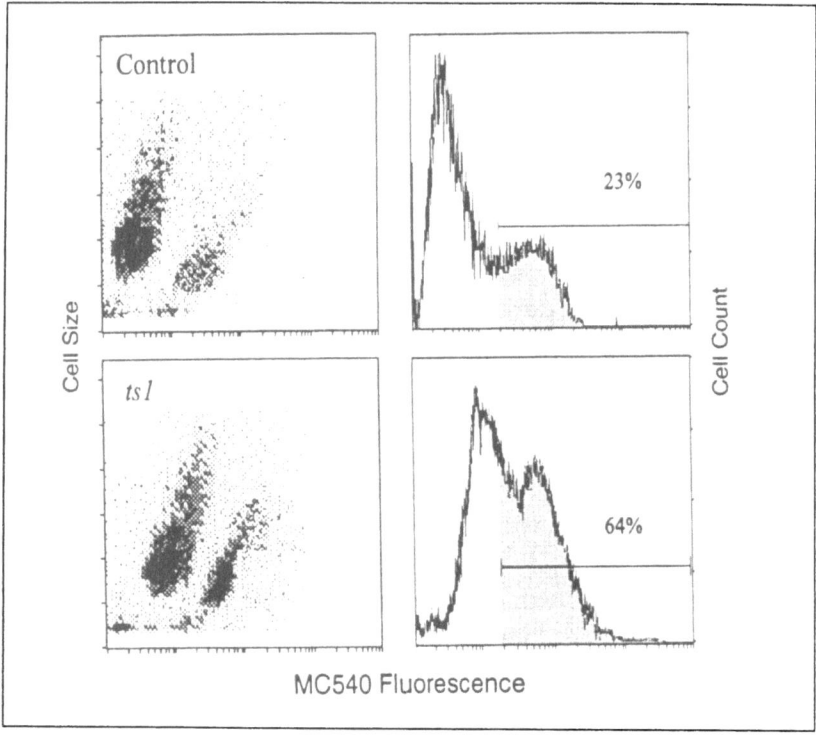

Fig. 4.4. Exposure of plasma membrane anionic phospholipids estimated using the affinity of MC540 for phosphatidylserine as indicator, in control and *ts*1 infected immortalized astrocytes at day 3 after infection. Dead and floating cells were removed before the fluorescence-activated cell sorter analysis. Twenty-three percent of the control cells bound MC540 tightly, whereas 64% of the infected cells bound MC540 tightly, and the remaining cells bound MC540 loosely; 70% of the control cells did not bind MC540.

beginning to shrink, vesiculate, and produce minicells containing fragmented DNA. This finding indicates that phosphatidylserine, which is normally distributed in the inner leaflet of the plasma membrane, became exposed early. Since these changes in membrane structure occurred quickly and preceded the excretion of vesicles containing fragmented DNA (Fig. 4.5), it appears that an early event in *ts*1 infection is to block the ATP-dependent transporter, which maintains the anionic phospholipids at their inner membrane site. But whether this inversion of plasma membrane phospholipids is directly due to binding of *ts*1 MoMuLV envelope to plasma membrane components is not established.

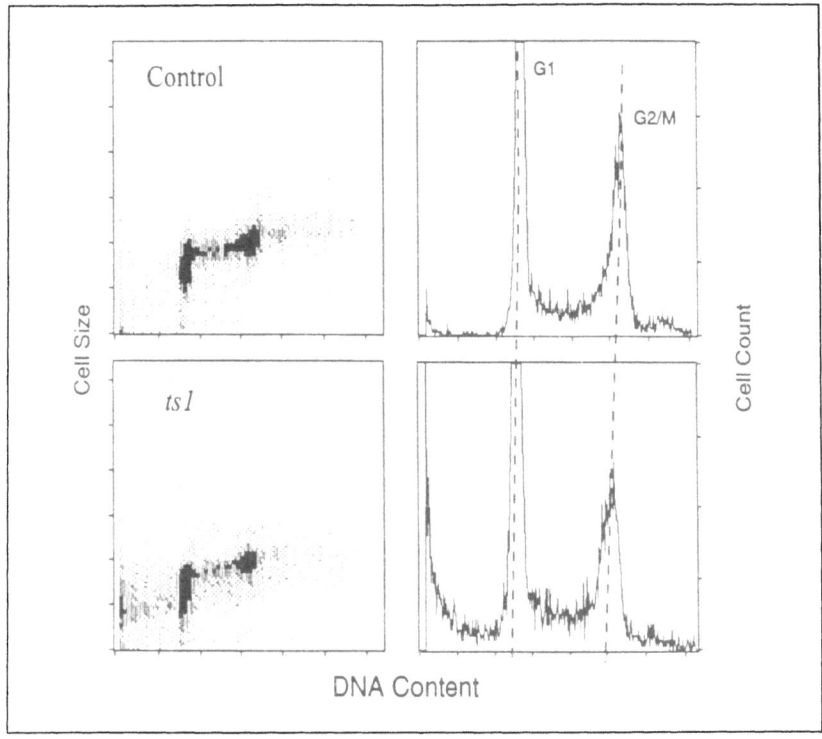

Fig. 4.5. Changes in DNA content and cell size caused by *ts1* MoMuLV in immortalized astrocytes 2 days after infection. DNA content was estimated in alcohol-fixed cells stained with propidium iodide by fluorescence-activated cell sorter analysis. Cell size was estimated by forward scatter. Many of the infected G2-M cells were converted into small vesicles containing sub-G1 amounts of DNA.

Glutamate Uptake

Uptake of excitotoxic glutamate is one of the many neuro-protective functions of astrocytic plasma membrane.[28] Recently, an ~25-30% reduction in glutamate uptake at 5 dpi and 10 dpi in *ts1*-infected primary astrocytes versus controls was observed (unpublished data). In comparison, the uptake of leucine did not change. This inhibitory effect of *ts1* MoMuLV was also observed in infected immortalized astrocyte cultures. However, whether this f:: ling re-flects what occurs in vivo is uncertain. Because astrocytes are responsible for maintaining low or optimal (nonexcitotoxic) concentrations of excitotoxic amino acids (EAAs) at the neuronal synapses, it is likely that this function of astrocytes would fail as astrocytes were forced to either die, differentiate, or proliferate due to the *ts1* infection. The notion that EAAs such as glutamate may play a neurotoxic role in *ts1*-infected CNS is substantiated by ultrastructural evidence that the degenerative changes within the neuropil in *ts1*-infected CNS occur first at the postsynaptic sites (dendrites and soma) and only later at the presynaptic terminals.[19] The above observations in *ts1* MoMuLV-infected astrocytes suggest that the *ts1* envelope peptides may, by binding to unknown components in the plasma membrane, alter the rates of the many signaling pathways involved in cell fate. Retroviral transmembrane envelope peptides obtained from MoMuLV or HIV[29,30] have been shown to greatly modulate the G protein- and cAMP-dependent transduction system at the plasma membrane.[29,30] It is possible, therefore, that a transmembrane segment of the *ts1* MoMuLV envelope protein may also activate either the cAMP-dependent pathways controlling cell fates at the plasma membrane or open the voltage-gated K^+ channel that activates the cell death pathways.[31]

Transgenic Studies

To determine whether the *ts1-env* gene product alone can cause neurodegeneration in vivo, we have generated several transgenic mouse founder lines harboring only the *ts1-env* gene with long terminal repeat of *ts1* or MoMuLV as the regulatory element.[10] These transgenic studies demonstrate that expression of *env* transcripts of *ts1* in the CNS of transgenic mice elicits neuropathological alterations that resemble the spongiform neurodegeneration observed following infection with replicating *ts1* virus, albeit the lesions seen were not as severe and took much longer to develop. Interestingly, transgenic mice expressing the CasBrE MuLV *env* gene[32] or HIV *env* gene[33] in the CNS also show a mild form of neurodegenerative lesions in the CNS. Thus, our data, together with those obtained with HIV and CasBrE MuLV, strongly support the notion that the retroviral *env* gene or its products is sufficient to induce neuropathogenic alterations in the CNS, resulting in neuronal degeneration.

Conclusions and Therapeutic Implications
in *ts1* MoMuLV-induced NID

If the "ER overload" hypothesis, which suggests overproduction of apoptosis-inducing signals such as Ca^{2+}, H_2O_2, ceramides, and GD3 gangliosides, proves to be correct, then many novel possibilities for therapeutic amelioration arise. To facilitate processing in

ER by induction of cysteine peptide isomerase-type chaperones (PDI or thioredoxins)[34-36] is one approach. To remove or buffer the excess Ca^{2+} and H_2O_2 produced by the clogged ER should also be beneficial in both astrocytes and T cells. The Ca^{2+} may also be removed either by activating the cytokine and cAMP-mediated Na^+/Ca^{2+} exchanger in the astrocytes[37,38] or by addition of intracellular Ca^{2+} binders.[39] The H_2O_2 can be buffered using many permeant antioxidants, including permeant thiols[40] and nitrone radical traps.[41] The proposed overactivity of cathepsin D can be suppressed by the protease inhibitor pepstatin A or by cathepsin D antisense RNA.[42] Since the cAMP-dependent pathways controlling some cell fate pathways may be homeostatically disrupted by the ts1 MoMuLV envelope peptides, inhibitors of adenylate cyclase[43] may be therapeutically useful in reestablishing homeostasis in the ts1 MoMuLV-induced unbalanced signaling pathways. The downstream caspase proteases responsible for mediating the apoptotic pathways may be blocked using specific caspase inhibitors.[44,45] If the above preventive measures can readjust the imbalanced cell fate pathways in the infected lymphocytes and/or the astrocytes, these virus altered cells may in time be able to readjust their own transcription-mediated defenses and differentiate into virus-resistant, quiescent cells that maintain their normal functions. However, with the persistent infection caused by integration of the retroviral genome into the host DNA, the surviving lymphocytes may with time become transformed, while the surviving astrocytes may continue to differentiate into activated macrophagelike cells and lose their neuronal support function. Our observations in the ts1 MoMuLV model of NID indicate that specific means for suppressing viral replication and preventing the cellular losses in this NID syndrome may be possible. Administration of either polyinosine cytosine[16] or antigen-primed adult cytotoxic CD8 T cells[46] to mice infected neonatally with ts1 was shown to prevent the cell death of both neurons and T cells as well as decrease viral titers, especially in the brain. But both reagents must be given early, before cellular death occurs and before the viral load becomes overwhelming.[16] Thus, by upregulating host defense systems in these neonatal mice, these agents can prevent the ts1-induced neuronal and T cell losses to a great extent. However, in time the surviving infected lymphoid cells in the thymus of the treated mice will become transformed, and the mice will die by 1 year of age of thymic lymphoma. Other means of boosting or homeostatically balancing the failing immune systems in these *ts*1 MoMuLV-infected neonatal mice have also been identified.[16] In vivo administration of redox agents (e.g., N-acetyl cysteine) plus the growth factors IL-2 or IL-4, were also shown to retard the cell losses. However, these treatments, unlike that with the above superantigen, could increase survival of the infected mice for only a few weeks. Another pathogenic possibility is that *ts*1 envelope either directly activates the protease caspase-3 or may activate down-stream the primary effector of an actin depolymerization, gelsolin, with collapse of the actin cytoskeleton. Our observations indicating that both lymphoid cells and astrocytes shrink, bleb, lose their ability to homotypically aggregate and excrete vesicles suggest that the actin cytoskeleton has collapsed in the infected cells. Kinetic measurements to confirm this K^+/cytoskeleton hypothesis have, however, not been done.

Acknowledgments

We thank C. McKinley and M. Lynn for their assistance in preparing the manuscript. We also thank Jude Richard for critical reading of the manuscript. This work is supported by Public Health Service grants CA45124 from the National Cancer Institute (NCI), AI28283 from the National Institute of Allergy and Infectious Diseases, MH57181 from the National Institute of Mental Health, Core Grant 16672 from the NCI, and a grant from the AT foundation, Austin, TX.

References

1. Wong PKY, Yuen PH. Molecular basis of neurologic disorders induced by a mutant, *ts*1, of Moloney murine leukemia virus, In: Roos RP ed. Molecular Neurovirology: Pathogenesis of Viral CNS Infection. New Jersey: Humana Press 1992:161-197.

2. Wong PKY, Russ LJ, McCarter JA. Rapid, selective procedure of isolation of spontaneous temperature-sensitive mutants of Moloney leukemia virus. Virology 1973; 51:424-431.

3. Wong PKY. Moloney murine leukemia virus temperature-sensitive mutants: A model for retrovirus-induced neurologic disorders. Curr Top Microbiol Immunol 1990; 160:29-60.

4. Gonzales-Scarano F, Nathanson N, Wong PKY. Retroviruses and the nervous system, In: Levy JA ed. The Retroviridae, vol. 4. New York: Plenum Press, 1995:409-490.

5. Szurek PF, Yuen PH, Ball JK, et al. A Val-25-to-Ile substitution in the envelope precursor polyprotein, gPr80[env], is responsible for the temperature sensitivity, inefficient processing of gPr80[env], and

neurovirulence of *ts1*, a mutant of Moloney murine leukemia virus TB. J Virol 1990; 64:467-475.

6. Kamps CA, Lin Y-C, Wong PKY. Oligomerization of the envelope protein of Moloney murine leukemia virus-TB and of *ts1*, a neurovirulent temperature-sensitive mutant of MoMuLV-TB. Virology 1991; 184:687-694.

7. Yu Y, Kamps CA, Yuen PH et al. Construction and characterization of expression systems for the env gene of *ts1*, a mutant of Moloney murine leukemia virus-TB. Virus Res 1991; 19:83-92.

8. Shikova E, Lin Y-C, Saha K et al. Astrocyte-specific defective gPr80^{env} processing correlates with cytopathogenicity induced by *ts1*, a mutant of Moloney murine leukemia virus. J Virol 1993; 67:1137-1147.

9. Wong PKY, Saha K, Lin Y-C et al. Long term cultivation and productive infection of primary thymocyte cultures by a thymocytopathic murine retrovirus. Virology 1996; 215:203-206.

10. Yu Y, Choe W, Stoica G et al. Development of pathological lesions in the central nervous system of transgenic mice expressing the *env* gene of *ts1* Moloney murine leukemia virus in the absence of the viral *gag* and *pol* genes and viral replication. NeuroVirology 1997; 3:274-282.

11. Wong PKY, Floyd E, Szurek PF. High susceptibility of FVB/N mice to the paralytic disease induced by *ts1*, a mutant of Moloney murine leukemia virus TB. Virology 1991; 180:365-371.

12. Stoica G, Floyd E, Illanes O et al. Temporal lymphoreticular changes caused by *ts1*, a paralytogenic mutant of Moloney murine leukemia virus-TB. Lab Invest 1992; 66:427-436.

13. Wong PKY, Prasad G, Hansen J et al. *ts1*, a mutant of Moloney murine leukemia virus-TB, causes both immunodeficiency and neurologic disorders in BALB/c mice. Virology 1989; 170:140-154.

14. Wong PKY, Szurek PF, Floyd E et al. Alteration from T to B cell tropism reduces thymic atrophy but not neurovirulence induced by *ts1*, a mutant of Moloney murine leukemia virus TB. Proc Natl Acad Sci USA 1991; 88:8991-8995.

15. Saha K, Yuen PH, Wong PKY. Murine retrovirus-induced destruction of CD4$^+$ T cells and thymocytes are mediated through activation-induced death by apoptosis. J Virol 1994; 68:2735-2740.

16. Lynn WS, Wong PKY. Neuroimmunopathogenesis of the ts1 MoMuLV viral infection. NeuroImmunoModulation 1998; in press.

17. Lynn WS, Wong PKY. Possible control of cell death pathways in ataxia telangiectasia—a case report. NeuroImmunoModulation 1997; 4:277-284.

18. Saha K, Wong PKY. *ts1*, a temperature sensitive mutant of Moloney murine leukemia virus-TB can infect both CD4$^+$ and CD8$^+$ T cells but requires only CD4$^+$ cells in order to cause paralysis and immunodeficiency. J Virol 1992; 66:2639-2646.

19. Stoica G, Illanes O, Tasca S et al. Temporal central and peripheral nervous system changes induced by a paralytogenic mutant of Moloney murine leukemia virus TB. Lab Invest 1993; 66:427-436.

19a. Chen TA, Morin PJ, Vogelstein B, Kinsler KW. Mechanisms under-
     lying nonsteroidal anti-inflammatory drug-mediated apoptosis. Proc
     Natl Acad Sci UXA 1998; 95:681-680.

20. Wong PKY, Yuen PH. Cell types in the central nervous system in-
    fected by murine retroviruses: Implications for the mechanisms of
    neurodegeneration. Histol Histopathol 1994; 9:845-848.

21. Lynn WS, Wong PKY. Neuroimmunodegeneration: Do neurons and
    T cells utilize common pathways for cell death? FASEB J 1995;
    9:1147-1156.

22. Wong PKY, Lynn WS. Neuroimmunodegeneration. EOS J Immunol
    Immunopharmacol 1997; 17:30-35.

23. Choe WK, Stoica G, Lynn WS et al. Neurodegeneration induced by
    MoMuLV-*ts1* and increased expression of TNFα and Fas in the cen-
    tral nervous system. Brain Res 1997; in press.

24. Zachary JF, Knupp C, Wong PKY. Non-inflammatory spongiform
    polioencephalomyelopathy caused by a neurotropic temperature-sen-
    sitive mutant of Moloney murine leukemia virus TB. Am J Pathol
    1986; 124:457.

25. Tsacopoulos M, Magestretti PJ. Metabolic coupling between glia and
    neurons. J Neurosci 1996; 16:877-885.

26. Lin YC, Chow CW, Yuen PH et al. Establishment and characteriza-
    tion of conditionally immortalized astrocytes to study their interac-
    tion with *ts1*, a neuropathogenic mutant of Moloney murine leuke-
    mia virus. Neuro Virol 1997; 3:28-37.

27. Benos DJ, Hahn BH, Bubien JK et al. Envelope glycoprotein gp120
    of human immunodeficiency virus type 1 alters ion transport in as-
    trocytes: Implications for AIDS dementia complex. Proc Natl Acad
    Sci USA 1994; 91:494-498.

28. Lipton SA, Rosenberg PA. Excitatory amino acids as a final com-
    mon pathway for neurologic disorders. N Engl J Med 1994;
    330:613-619.

29. Haraguchi S, Good RA, James-Yarish M et al. Induction of intracel-
    lular cAMP by a synthetic retroviral envelope peptide: A possible
    mechanism of immunopathogenesis in retroviral infections. Proc
    Natl Acad Sci USA 1995; 92:5568-5571.

30. Levi G, Patrizio M, Bernardo A et al. Human immunodeficiency vi-
    rus coat protein gp120 inhibits the β-adrenergic regulation of
    astroglial and microglial functions. Proc Natl Acad Sci USA 1993;
    90:1541-1545.

31. Yu SP, Yeh C-H, Sensi SL et al. Mediation of neuronal apoptosis by
    enhancement of outward potassium current. Science 1997; 278:114-117.

32. Kay DG, Gravel C, Pothier F et al. Neurological disease induced in
    transgenic mice expressing the *env* gene of the Cas-Br-E murine
    retrovirus. Proc Natl Acad Sci USA 1993; 90:4538-4542.

33. Toggas SM, Masliah E, Rockenstein EM et al. Central nervous sys-
    tem damage produced by expression of the HIV-1 coat protein gp120
    in transgenic mice. Nature 1994; 367:188-193.

34. Ezzell C. Protein folding and the early secretory pathway: Researchers begin to understand the cellular assembly line. J NIH Res 1997; 9:42-47.

35. Walker KW, Gilbert HF. Scanning and escape during protein-disulfide isomerase-assisted protein folding. J Biol Chem 1997; 272: 8845-8848.

36. Gottesman S, Wickner S, Maurizi MR. Protein quality control: Triage by chaperones and proteases. Genes Dev 1997; 11:815-823.

37. Kirischuk S, Verkhratsky A. Calcium homeostasis in aged neurons. Live Sciences 1996; 59:451-459.

38. Kirischuk S, Kettenmann H, Verkhratsky A. NA$^+$/CA2$^+$ exchanger modulates kainate-triggered CA2$^+$ signaling in Bergmann glial cells in situ. FASEB J 1997; 11:566-572.

39. Chiou CY, Malagodi MH. Studies on the mechanism of action of a new Ca^{2+} antagonist 8-(N.N dithylamino) octyl 3,4,5-trimethoxybenzoate hydrochloride in smooth and skeletal muscles. Br J Pharmacol 1995; 53:279-285.

40. Sandstrom PA, Mannie MD, Buttke TM. Inhibition of activation-induced death in T cell hybridomas by thiol antioxidants: Oxidative stress as a mediator of apoptosis. J Leuk Biol 1994; 55:221-229.

41. Floyd RA, Carney JM. Nitrone Radical Traps Protect In Experimental Neurodegenerative Diseases, Free Radical Biology & Aging Research. Oklahoma City, OK: Academic Press Limited, 1996:70-83.

42. Deiss LP, Galinka H, Berissi H et al. Cathepsin D protease mediates programmed cell death induced by interferon-γ, Fas/APO-1 and TNF-α. EMBO J 1996; 15:3861-3870.

43. Haraguchi S, Good RA, Cianciolo GJ et al. Immunosuppressive retroviral peptides: Immunopathological implications for immunosuppressive influences of retroviral infections. J Leukoc Biol 1997; 61:654-666.

44. Saftig P, Hetman M, Schmahl W et al. Mice deficient for the lysosomal proteinase cathepsin D exhibit progressive atrophy of the intestinal mucosa and profound destruction of lymphoid cells. EMBO J 1995; 14:3599-3608.

45. Mansat V, Bettaieb A, Levade T et al. Serine protease inhibitors block neutral sphingomyelinase activation, ceramide generation, and apoptosis triggered by daunorubicin. FASEB J 1997; 11:695-702.

46. Saha K, Wong PKY. Protective role of cytotoxic lymphocytes against murine leukemia virus-induced neurologic disease and immunodeficiency is enhanced by the presence of helper T cells. Virology 1992; 188:921-925.

CHAPTER 5

LP-BM5 MuLV Infection: Impact on the Immune and Central Nervous Systems

Michael G. Espey, Yelena Kustova, Yoshitatsu Sei, and Anthony S. Basile

Introduction

Mice infected with the LP-BM5 retrovirus mixture develop a syndrome of lymphoproliferation and immunodeficiency similar to the acquired immunodeficiency syndrome (AIDS) developed in patients infected with HIV-1 (Table 5.1).[1-3] Although the primary target of both HIV-1 and LP-BM5 is the immune system, both viruses elicit secondary effects on the central nervous system (CNS) during the course of infection. This chapter presents an overview of the characteristics of LP-BM5 infection as it relates to the development of a syndrome of neuronal dysfunction that parallels many of the features of AIDS dementia complex (ADC), particularly in its early stages.

Characteristics of LP-BM5-Induced Immunodeficiency

LP-BM5 contains a mixture of C-type murine leukemia viruses which were originally derived from C57BL/6 (B6) mice that had been inoculated with cell-free extracts of radiation-induced thymomas.[4,5] The primary pathologic agent in LP-BM5 is a replication-defective virus termed BM5d (also referred to as DuH5). The mixture also contains replication-competent B-tropic ecotropic and amphotropic mink cell focus-inducing viruses.[1,6] Coinfection with ecotropic or amphotropic helper viruses potentiates the development of disease

Neuroimmunodegeneration, edited by Paul K.Y. Wong and William S. Lynn.
© 1998 Springer-Verlag and R.G. Landes Company.

Table 5.1. Comparison of the characteristics of LP-BM5 and HIV-1 infections

SIMILARITIES

Immune System
CD4$^+$, CD8$^+$ lymphocyte and NK cell anergy.
Hyporesponsiveness to infectious agents and antigenic stimulation.
Progressive increase in plasma cytokine levels: IL-4, 10, IFN-γ.
Polyclonal B lymphocyte activation.
Hypergammaglobulinemia.
Autoimmune phenomena.
Development of lymphoma with extralymphoid metastasis and paraneoplasia.
Enhanced susceptibility to secondary infections.
Activation-induced apoptosis.

Central Nervous System
Non-neuronotropic.
Development of learning and memory deficits.
Global microglial activation with nodule formation.
Astrocytosis with aggregation.
Selective defects in blood-brain barrier integrity.
Pleocytosis of choroid plexus and meninges.
Neurotoxin production: Glutamic acid; PAF; quinolinic acid; TNF-α.
Primary site of lesion: Basal ganglia.

Other
Wasting syndrome.
Alterations in tryptophan metabolism
Influence of age and MHC polymorphism on disease progression.

DIFFERENCES

LP-BM5 is a C-type murine leukemia virus; HIV-1 is a human lentivirus.
LP-BM5 has wider cell tropism (T, B lymphocytes, microglia, astrocytes, vascular endothelium) than HIV-1.
Initial target of LP-BM5 is B lymphocyte; CD4$^+$ T lymphocyte for HIV-1.
Kaposi's sarcoma is seen with HIV-1 infection. Strain dependent lymphadenopathy and splenomegaly result from LP-BM5 infection.
Myelin pallor, macrophage infiltrates, and multi-nucleated giant cells in brain parenchyma in association with HIV-1 infection. Some evidence of neuronal death in late stages.

by facilitating the formation and spread of pseudotyped BM5d genomes.[7] The BM5d genome encodes a 60 kDa Gag fusion precursor protein (Pr60^{gag}) which differs markedly from the nonpathogenic ecotropic virus in the p12 and p15 regions.[7] Myristylation and insertion of the Pr60^{gag} into the plasma membrane is a critical determinant of disease induction.[8,9] $Pr60^{gag}$ can also bind to c-Abl, a tyrosine kinase with oncogenic potential, increasing its translocation from the nucleus to the cell surface.[10]

The presence or absence of BM5d in spleen extract 3 to 12 months after inoculation with LP-BM5 correlates with disease susceptibility or resistance among inbred strains of mice.[11-13] The most thoroughly studied strain is B6, in which intraperitoneal inoculation with LP-BM5 induces a syndrome termed murine AIDS (MAIDS) with a duration of 16 to 18 weeks. The most striking presentation of LP-BM5 infection is the progressive enlargement of the spleen and lymph nodes. It is not uncommon to observe spleen weights in excess of 2,500 mg (normal ≈ 65mg) in end-stage animals. In addition to splenomegaly and lymphadenopathy, MAIDS is characterized by functional and phenotypic changes in leukocyte populations (Table 5.1).[3,14] Polyclonal activation of infected B lymphocytes is observed, leading to oligoclonal B cell expansion and hypergammaglobulinemia. Effector cells (e.g., macrophages, NK cells) and T lymphocytes become anergic, failing to respond to mitogenic stimulation and orchestrate immune responses. Emergence of lymphoma is common in MAIDS,[15] due to either BM5d-induced transformation (e.g., stimulation of the protooncogene $v\text{-}abl^{10}$) or loss of tumor suppression capacity secondary to the development of anergy. A complex pattern of cytokine expression in peripheral lymphoid cells is also observed. This includes elevations in cytokines associated with humoral immunity: interleukin (IL)-4, IL-6, IL-10; and cytokines associated with cell mediated immunity: IL-1β, IL-2, IL-12, tumor necrosis factor-α and interferon-γ.[3,14]

The induction and progression of MAIDS not only depends upon virus infection, but also upon a cascade of interactions between cells of the immune system. Both B and $CD4^+$ T lymphocytes are required for the induction of disease.[16,17] Mature B lymphocytes are the primary targets of LP-BM5 and are required for efficient infection of T lymphocytes and macrophages.[18] BM5d is not expressed in B lymphocyte-deficient mice infected with LP-BM5, despite the persistence of ecotropic helper virus and integrated proviral BM5d DNA. Moreover, T lymphocyte-deficient strains of mice infected with LP-BM5 express high levels of BM5d, but fail to develop MAIDS.[19] Two hypotheses may help to explain the pathogenic mechanisms leading to the development of MAIDS. The first posits that disease may be driven by the presence of $Pr60^{gag}$ on the surface of infected

B lymphocytes. This acts as a superantigen capable of aberrantly stimulating T lymphocytes and leads to a degenerative cycle of T and B cell proliferation, cytokine production and immunodeficiency.[3,20] Alternatively, disease may be related to the capacity of BM5d to induce oncogenic transformation of infected cells, leading to a paraneoplastic syndrome of immunodeficiency.[2,10]

LP-BM5 Associated Neuropathology

Behavioral Changes

The first evidence of a neuropathology in LP-BM5 infected mice was the demonstration by Sei et al[21] of selective deficits in spatial learning and memory using the modified Morris water maze. In this paradigm, mice are placed into a tank filled with water opacified with powdered milk.[22] Mice are allowed to swim until they locate a submerged platform. During 8 days of place-navigation training, the mice learn to find the platform using visual cues located in the room. Mice infected with LP-BM5 develop a spatial learning deficit between 8-10 wks postinoculation (PI). This is manifested as the inability to display either the target-directed search pattern or spatial preference characteristics of uninfected littermate control mice (Fig. 5.1). In contrast with control mice, infected mice require more training trials before showing a decrease in the time spent searching for the target, signifying a learning impairment. Moreover, in quadrant preference trials (where the target is removed), infected mice show no quadrant preference, suggesting that they have a poor memory of target location. LP-BM5 infected mice display no deficits in finding a visible platform, and show normal rotarod performance, suggesting that visual and motor function are normal. The cognitive deficits manifested by these mice are associated with peripheral LP-BM5 infection and the induction of MAIDS, because mice infected with ecotropic virus alone do not develop behavioral deficits, nor do LP-BM5 infected mice treated with azidothymidine, an anti-retroviral agent which reduces LP-BM5 titer and the severity of MAIDS.[23]

Virus Presence in the Brain

Following the report that LP-BM5 infected mice develop cognitive deficits, investigations of the CNS pathology contributing to these behavioral abnormalities commenced. No evidence of viral infection of neurons has been observed under any conditions, either in vitro or in situ following intraperitoneal or intracisternal in-

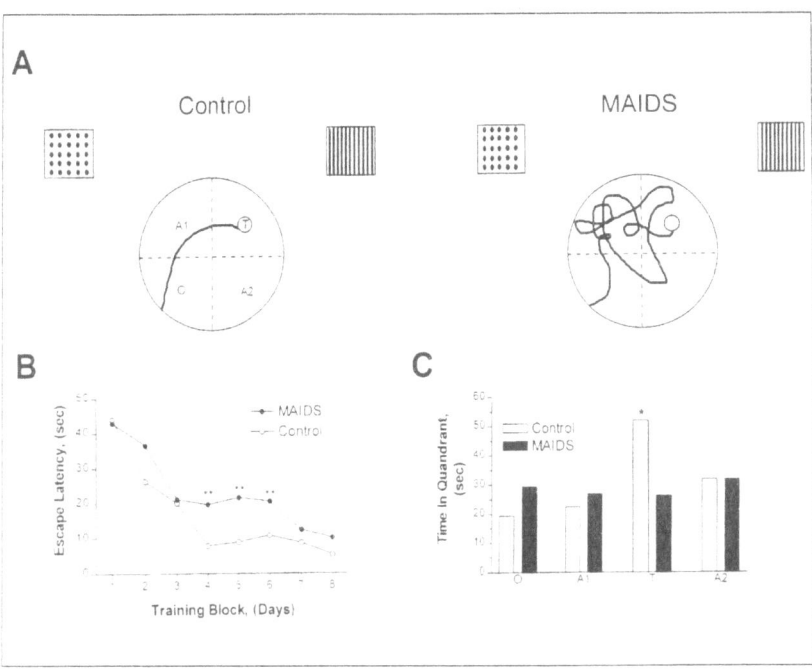

Fig. 5.1. Mice infected with the LP-BM5 MuLV show performance deficits in the Morris water maze. An overhead view of the swim tank used for testing is shown in panel (A). The tank is divided into four quadrants: the target quadrant (T) where a stable platform is located 1 cm below the surface of the opacified water; the two quadrants adjacent to the target quadrant (A1 and A2); and the quadrant opposite the target (O), where the mouse begins its swim. Visual cues (the squares containing dots and lines) are placed on the walls around the tank, and are used by the mouse for spatial orientation. Typically, mice will undergo 16 trials (two per day, spaced over 8 days) consisting of swims lasting approximately 2 minutes. The representative paths taken by control (left) and LP-BM5 infected mice (right) in the swim tank at the 12th trial are shown. While the control mouse swims directly to the platform, the mouse infected for 10 weeks with LP-BM5 follows a circuitous route to the target. The deficit in the rate of spatial learning of LP-BM5 infected mice is indicated in panel B. While control mice reach the learning criterion (≈10 seconds to reach target) after 8 trials (4 days), LP-BM5 infected mice require more trials (14) to reach criterion. Spatial memory is tested by the quadrant preference paradigm (Panel C). After mice reach the criterion for escape latency (finding the target in ≈10 seconds), the target is removed. The amount of time spent in the original target quadrant upon retesting is indicative of the residual memory the animal holds for the spatial location of the target. Control mice preferentially spend significantly more of their time in the target quadrant than the other quadrants (P<0.05, Univariate F-test). In contrast, LP-BM5 infected mice show no quadrant preference, suggesting that they have a poor memory of the spatial location of the target.

oculations. However, astrocyte enriched cultures of neonatal glia can be productively infected with LP-BM5.[24] During the typical course of infection of adult B6 mice, there is a paucity of virus or viral proteins in the brain parenchyma, particularly in comparison to peripheral lymphoid tissues. However, some evidence for the presence of Pr60[gag] and ecotropic virus Gag has been found in the CNS of infected mice using immunohistochemistry. Immunoreactivity for these two viral proteins are localized in cells of the subventricular zone and in macrophage-like cells associated with the choroid plexus (epiplexus), meninges, and ependyma as early as 2 weeks PI. Presently, it is not known whether these cells are productively infected or has sequestered virus or Pr60[gag] into phagocytic vesicles. Using RT-PCR, very low levels of BM5d mRNA have been detected in whole brain, consistent with a minimal, productive brain infection.[25] Thus, while there is some evidence for the presence of virus in the CNS during the typical course of LP-BM5 infection of B6 mice, the levels of virus message and protein are low and restricted primarily to cells in the meninges, ventricular surfaces and the subventricular zone.

Infiltration of the CNS by leukocytes and monocytes is a common feature of HIV-1 infection, with perivascular cuffs and multinucleated giant cells figuring prominently in this pathology.[26] Moreover, infiltration of infected monocytes into the brain parenchyma via perivascular spaces, as well as leukocytic traffic through the choroid plexus,[27,28] may serve as significant routes by which HIV-1 enters the brain. In contrast, the leukoencephalitic response to LP-BM5 infection is primarily limited to the leptomeninges and choroid plexus, and correlates well with the localization of the small amounts of virus present. Pleocytosis is prevalent only in the latter stages of MAIDS (≥12 weeks PI) when MHC II immunoreactive lymphocytes and macrophages are observed (Fig. 5.2A,B). Malignant transformation of lymphocytes in hyperplastic peripheral compartments often occurs during this period of infection. With progression of disease, these cells metastasize into nonlymphoid tissues, including the periventricular regions of the brain.[15,29] Perivascular cuffing and egress by leukocytes into the neuropil are not salient features of infection in B6 mice. This is noteworthy since expression of the leukocyte adhesion molecules ICAM and VCAM is increased on both the endothelia of cerebral vasculature, choroid plexus and meninges. The lack of leukocytic diapedesis across the vascular walls and into the CNS parenchyma during LP-BM5 infection may be related to MAIDS-induced leukocyte anergy.[2,3,30]

In summary, little virus is found in the brain during the typical course of LP-BM5 infection of B6 mice despite profound peripheral immune system disease. Moreover, while ventricular structures may serve as the main route of entry for those few virus-laden leukocytes that appear in the CNS, a cytopathic infection of mature neu-

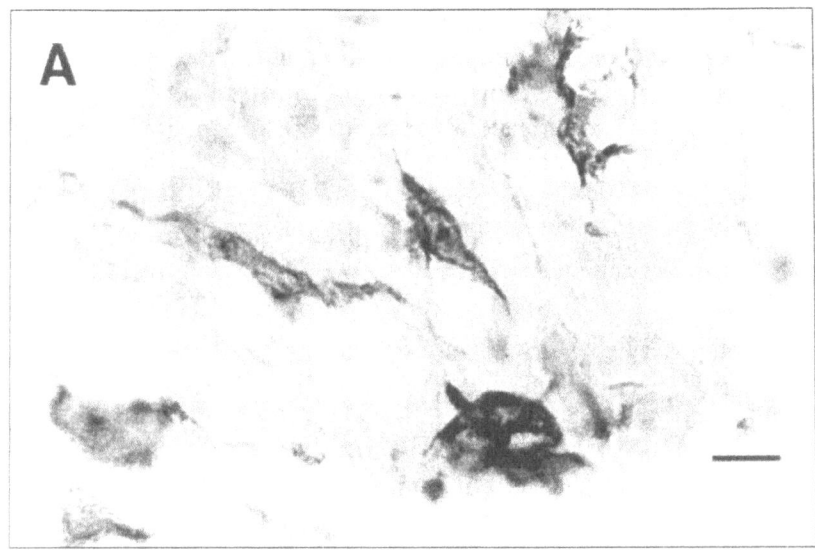

Fig. 5.2. Choroid plexus in LP-BM5 infected mice. (A) Epiplexus cells strongly express MHC II antigens (Ia). Magnification = 300 x. Bar = 35 μm. (B) Quinolinic acid immunoreactive macrophages and lymphocytes are visible in the fibrous base to which the choroid plexus is attached. Magnification = 1500 x. Bar = 7 μm.

rons does not occur. Given a longer survival time resulting from increased resistance to infection (e.g., employing a different mouse strain),[11-13] viral invasion of the CNS by LP-BM5 infected monocytes may occur to a greater extent than observed in B6 mice. The presence of high densities of infected monocytes in the brains of these mice may give rise to additional pathological processes than those observed thus far. However, it is debatable whether lymphoid cells in the CNS would remain immunocompetent and mediate a leukoencephalitic response in the late stages of MAIDS when peripheral immune responsiveness is profoundly deficient. Thus, the neuropathologies contributing to the spatial learning deficits observed in LP-BM5 infected mice most likely arise through indirect mechanisms independent of the presence of virus in the brain.

Gliosis Is a Prominent Early Feature of LP-BM5 Infection

Despite the paucity of virus within brain parenchyma, prominent changes in glial activation develop soon after infection with LP-BM5.[31] A 5-fold increase in F4/80 (a marker of microglia/macrophage activation) expression was evident by 4 weeks PI and was sustained in all regions tested (except hippocampus) for 12 weeks. F4/80 immunoreactivity was clearly associated with hypertrophic microglia, the endogenous macrophage of the CNS (Fig. 5.3). Moreover, a distinct class of microglia (15-20% of the total F4/80-positive cells) with the characteristics of microglial nodules was evident in the striatum and corpus callosum of mid-stage mice. Microglia with an ameboid or foamy macrophage morphology, a phenotype indicative of neuronal death, were absent.

Increased GFAP immunoreactivity was visible in the cortex and striatum at 5 weeks PI, but returned to control levels by 10 weeks.[31] Significant astrocytic hypertrophy was also observed, with increases in overall size and number of processes (Fig. 5.4). Activated astrocytes were not uniformly distributed throughout the cortex and striatum, but tended to aggregate, suggesting their association with lesion foci. In contrast with F4/80 expression, the increase in GFAP immunoreactivity was not global, nor was it sustained. The decrease in GFAP expression in the later stages of disease may result from continued activation of astrocytes,[32] possibly in response to the presence of cytokines such as TNF.[33] It is important to emphasize that glial activation is one of the earliest events following infection with LP-BM5 infection, and occurs without evidence of either overt viral presence or infected leukocytic infiltration into the parenchyma. These observations raise the possibility that glial activation results from soluble substances (e.g., cytokines, viral proteins) released by lymphoid cells either in the choroid plexus or the periphery. The subsequent release of neurotoxic agents by activated microglia[34] and astrocytes suggests that the cognitive deficits observed in these mice result from neuronal responses to these toxic agents.

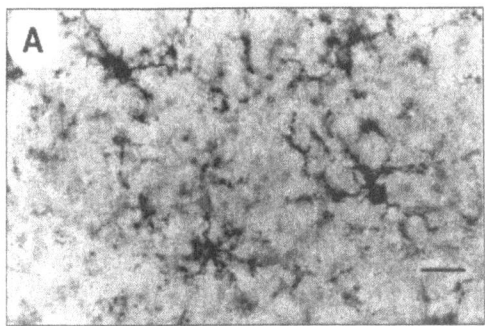

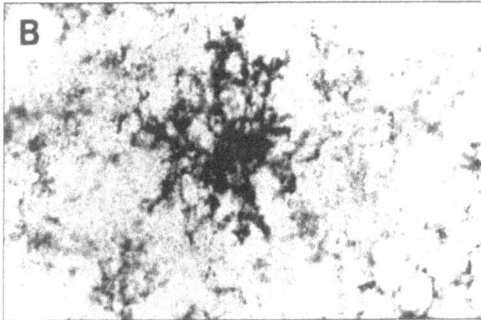

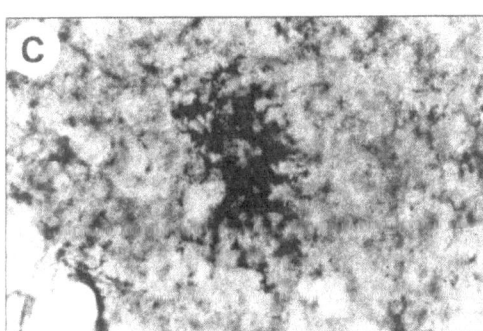

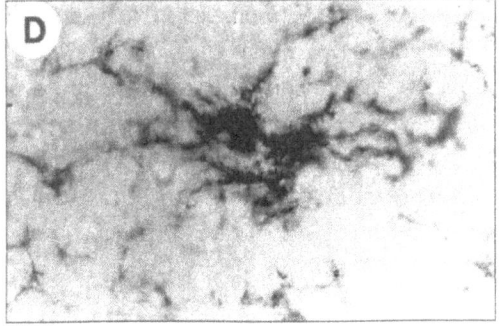

Fig. 5.3. Activated microglial clusters in LP-BM5 infected mouse brain regions. Sections are stained with antibodies to F4/80 protein. Typical resting microglia are shown in the striatum of normal C57BL/6 mice (A). Note the low intensity of staining, relatively few ramifications and the relatively large distance between cells. In the striatum of LP-BM5 infected mice (B,C), activated microglia form nodules of intensely stained cells with large numbers of finely ramified projections. Microglial clusters of somewhat different morphology are evident in the corpus callosum in infected mice (D). Magnification on all panels = 900 x. Bar = 12 μm.

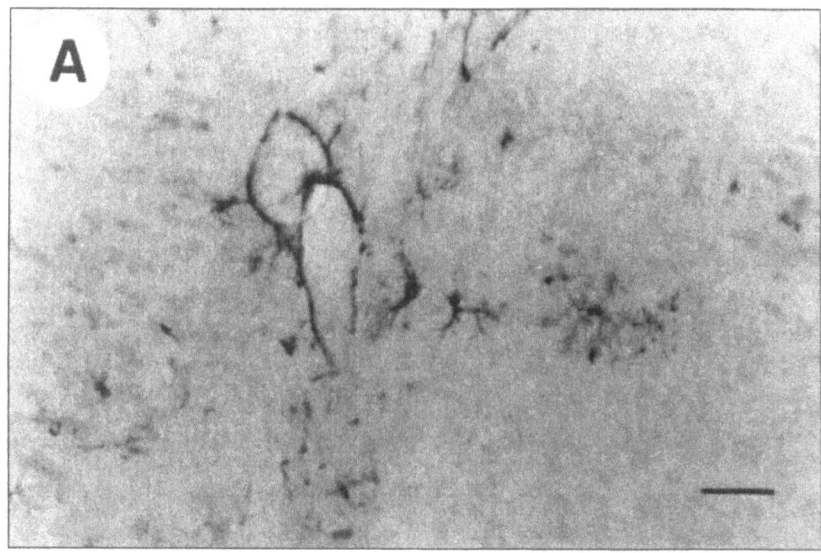

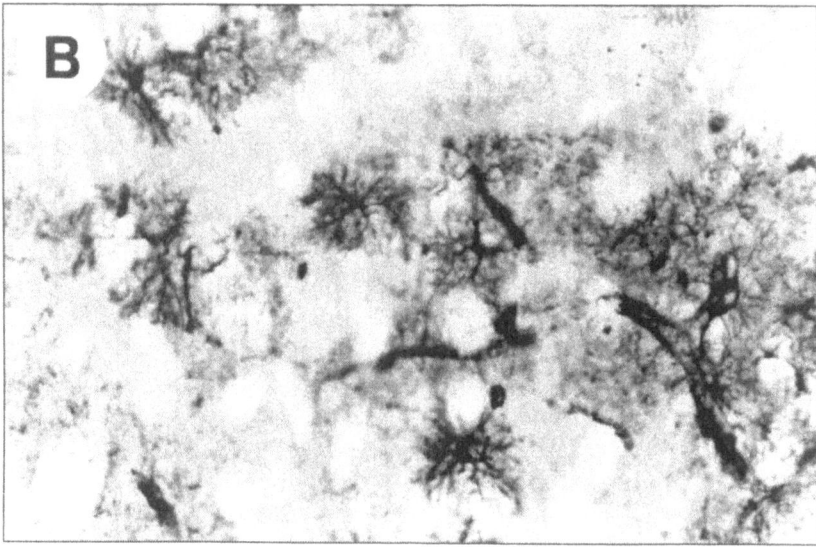

Fig. 5.4. Activated astrocytes in the striatum of LP-BM5 infected mice. Thick brain sections (50 μm) are stained with antibodies to GFAP. Brains from control mice show few areas with GFAP positive cells (A). Those regions with GFAP positive cells contain astrocytes of relatively small size, with few branches. In contrast, brains from LP-BM5 infected mice, fixed and stained under the same conditions, contain large numbers of intensely immunopositive astrocytes. These activated astrocytes are extensively arborized and tend to aggregate. Magnification on all panels = 600 x. Bar = 16 μm.

Neurotoxins and Their Effect on Neurotransmitter Systems

Insights into the nature of the neurotoxin(s) secreted or allowed to accumulate by the activated glia are provided by the pattern of neurochemical alterations which occur during the course of LP-BM5 infection. Throughout the brain of LP-BM5 infected mice, the concentrations of acetylcholine, met-enkephalin and substance-P are depleted, with the greatest decreases observed in the striatum and hypothalamus as early as 4 weeks PI.[35] Decreases in cNOS protein and mRNA expression were also observed, particularly in the striatum.[36] Although trifluoperazine-sensitive nitric oxide synthase activity was transiently increased in the cerebral cortex, striatum and cerebellum from 4-8 weeks PI, this probably represents an increase in astrocytic eNOS expression. Although the status of neurotransmitter receptor expression has not been extensively investigated in LP-BM5 infected mice, significant alterations in AMPA receptor expression were observed, as was previously reported in cerebella from patients with ADC.[37] This was manifested as a 30-50% decrease in the density of [³H]AMPA binding throughout the brain, beginning as early as 8 weeks PI (Fig. 5.5).[38] Moreover, this depletion was manifested in the cerebellum as a specific decrease in the expression of GluR-3 subunits by Purkinje neurons.

In addition to the neurochemical changes, significant structural changes consistent with an excitotoxic lesion are found. Bodian silver staining of the cerebral cortex from LP-BM5 infected mice shows scattered foci of reduced neurofiber staining. More selective analysis of MAP-2, and NF-200 expression indicated a 30-80% decrease in the immunoreactivity of these structural proteins in regions of the cerebral cortex and striatum, as indicated by immunoblotting and histochemistry. The immunoreactivity of synaptophysin was impacted to a lesser degree (25-57%) and in fewer brain regions than the other postsynaptic structural proteins. The function of the protein tyrosine kinase Fyn was also altered in the hippocampus of infected mice.[39] Among its many functions, Fyn helps to regulate neurofilament assembly. It is also the only "nonreceptor" type of protein tyrosine kinase (including Src and Yes) which is involved in long term potentiation and spatial memory.[40] Fyn kinase was maximally activated in hippocampal slices from LP-BM5 infected mice, as indicated by its inability to phosphorylate enolase or to undergo

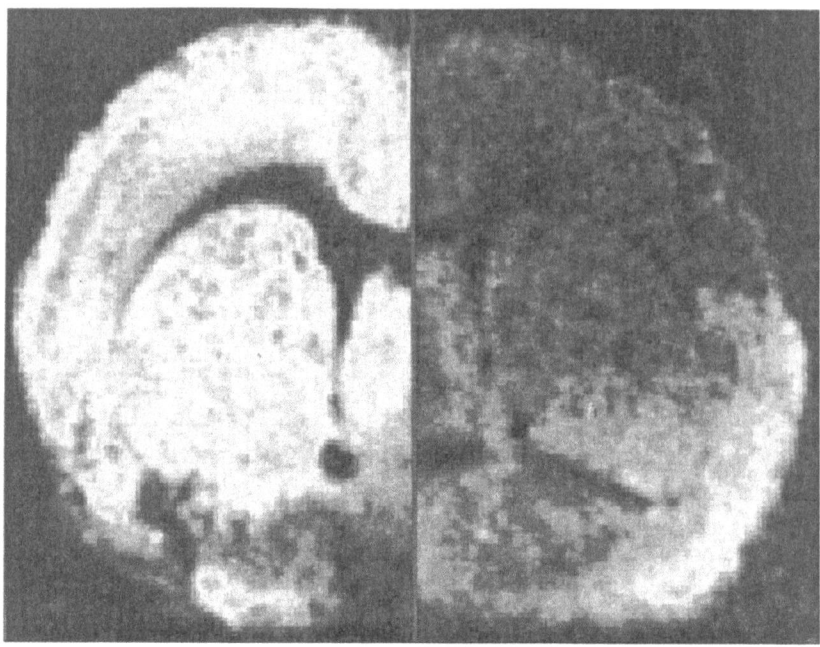

Fig. 5.5. False color coded autoradiograph of [³H]AMPA (30 nM) binding to coronal sections of brain from C57BL/6 mice infected with the LP-BM5 virus for 16 weeks (right panel), and normal C57BL/6 mice. High binding density is indicated by red, low binding density by blue. The density of [³H]AMPA binding was converted from grey scale values to nCi/mg tissue for each section using ³H microscales. [³H]AMPA binding to the somatosensory/motor cortex, lateral septal nucleus, and caudate/putamen from the LP-BM5 infected mouse was 85, 59, and 60% below control values, respectively. Binding density to the piriform cortex was unchanged. For color representation see page 151 in Color Insert.

autophosphorylation in response to stimulation by glutamate. Moreover, the distribution of Fyn immunoreactivity within pyramidal neurons was highly abnormal. Instead of being evenly distributed along the soma and dendrites, Fyn was concentrated in the cell bodies. Together, these results suggest that the proteins supporting the postsynaptic structures of neurons, particularly in the cortex, striatum and hippocampus, are significantly changed, with evidence of secondary degradation of presynaptic structures as well. This dendritic pathology is also observed in the cortex of humans infected with HIV-1[41] and mice infected with neurotropic retroviruses (e.g., Moloney *ts*1).[42] Degeneration of these postsynaptic structures is consistent with their relatively high density of glutamate receptors, making them vulnerable to excitotoxic insult. These changes may cause defects in dendritic spines and synaptic deafferentation, which may impair learning and memory processes.

These data suggest that a chronic, low level neurodegenerative process is occurring in the brains of LP-BM5 infected mice that may contribute to the development of their spatial learning and memory deficits. However, the identity of the toxic agent(s) responsible for this damage is only now becoming clear. Toxic levels of NO can be achieved in the brain by the induction of nitric oxide synthase (iNOS) in infiltrating macrophages, and has been implicated in neuron damage associated with HIV-1 infections.[43] While large numbers of iNOS immunopositive cells are found in the spleen, there is no evidence for the presence of iNOS message, protein, or enzyme activity in the brains of LP-BM5 infected mice up to 12 weeks post inoculation.[36] Platelet activating factor (PAF) is a potent proinflammatory neurotoxin whose levels are significantly elevated in the cerebrospinal fluid of individuals infected with HIV.[44] PAF concentrations are also increased throughout the brain of mice as early as 6 weeks after LP-BM5 inoculation.[45] The PAF found in the CNS of infected mice most likely originated from cellular elements in the brain, since PAF levels in the spleen were not significantly elevated above control. Moreover, PAF levels appeared to be regulated by glutamate, since treating infected mice with NMDA receptor antagonists reduced brain concentrations of PAF. A third putative neurotoxin investigated in LP-BM5 infected mice is quinolinic acid, which is an NMDA receptor agonist and excitotoxin.[47] Brain quinolinic acid levels are significantly increased at 12 weeks PI.[48] Leukocytes in peripheral lymphoid tissues and the choroid plexus[29] are probably the source of most or all of the quinolinic acid in the brains of LP-BM5 infected mice (Fig. 5.2B), which then gains entry to the CNS across a permeabilized blood-brain barrier. Regardless, the excitotoxic actions of quinolinic acid may not play an important role in producing the neurodegeneration that contributes to the cognitive deficits in LP-BM5 infected mice. This is due to the low concentrations of quinolinic acid achieved in the whole brain (≈ 1 μM) relative to its affinity for the NMDA receptor ($EC_{50} \approx 2.3$ mM).[49] Moreover, the decrease in AMPA receptor density in the brains of infected mice is not a direct response to receptor occupation by quinolinic acid, which has no measurable affinity for this glutamate receptor subtype. Finally, the temporal correlation between the elevations of quinolinic acid in the lymphoid tissues and plasma (and subsequently the CNS) and the development of spatial learning deficits is poor.[29,48]

Quinolinic acid and PAF are not the only potential neurotoxic agents elevated in the brains of LP-BM5 infected mice. Analysis of the cerebrospinal fluid from LP-BM5 infected mice revealed a significant, 40% increase in the concentration of the excitatory amino acid neurotransmitter glutamate (Fig. 5.6A). This concentration of glutamate in the CSF is within the range of that reported for patients suffering from ischemic strokes.[50] Moreover, the plasma concentration of glutamate was not altered by virus infection (Fig. 5.6B), and the gradient for other amino acids (e.g., alanine, arginine, serine) was maintained, suggesting that the source of glutamate in the CSF is not of extracerebral origin. Amino acids were also sampled from the extracellular milieu of the brain parenchyma using microdialysis. The steady-state levels of glutamate in the striatum were found to be increased by an average of ≈300% (Fig. 5.6C). These free glutamate concentrations are similar to those reported in animal models of focal or global ischemia.[51] The free glutamate concentrations in the extracellular fluid of infected mice may arise from both hyperactivated neurons and non-neuronal sources, such as activated microglia.[34] In addition, free glutamate levels may be elevated as a result of dysfunctional glutamate uptake. Preliminary analysis of glutamate transporter proteins suggest that both the neuronal (EAAC-1) and astroglial (GLAST) types of transporter are altered in the striatum of infected mice, with a decrease in EAAC-1 and an increase in GLAST. The increase in GLAST expression may be a response to increased free glutamate levels, possibly resulting from the suppression of pump function by agents such as arachidonic acid.[52] In contrast, the decrease in EAAC-1 expression may reflect the degeneration of neuronal terminals where the transporter is located.

Fig. 5.6. (opposite) Glutamate concentrations in brain fluids and plasma from 1-29 mice infected for 8 weeks with LP-BM5. CSF glutamate concentrations (A) were significantly increased (P<0.05, Kruskal-Wallis test) above control (n=12) values in infected mice. However, there was no change in plasma glutamate levels with infection, and the concentrations were lower than those observed in the CSF (B). Moreover, the concentration of glutamate in striatal microdialysates from infected mice is 3-fold higher than controls (n=18). These results suggest that the glutamate measured in the CSF and microdialysates from infected mice is of intracerebral origin. Possible sources of this glutamate include hyperactivated neurons, direct release from activated microglia, and/or accumulation as a result of impaired astrocytic or neuronal glutamate uptake.

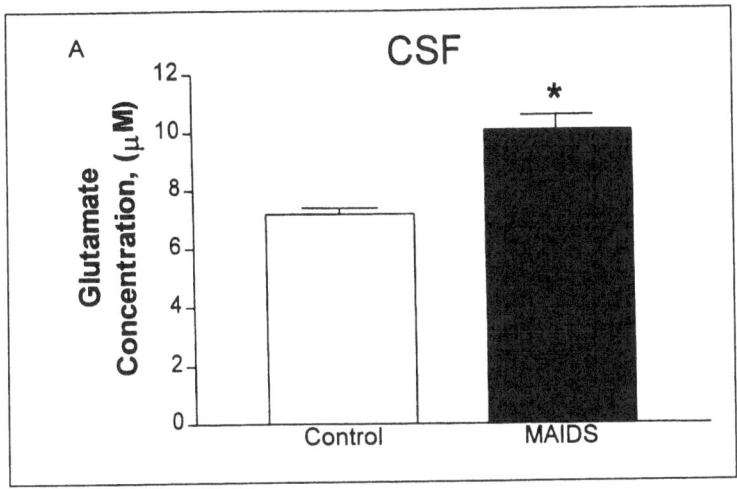

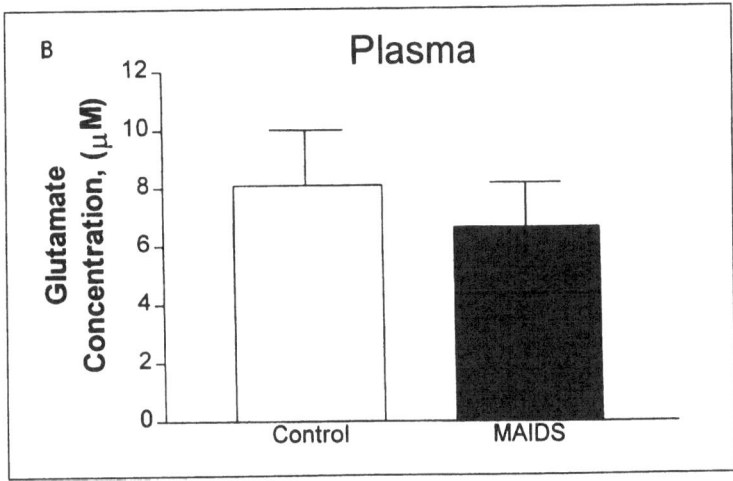

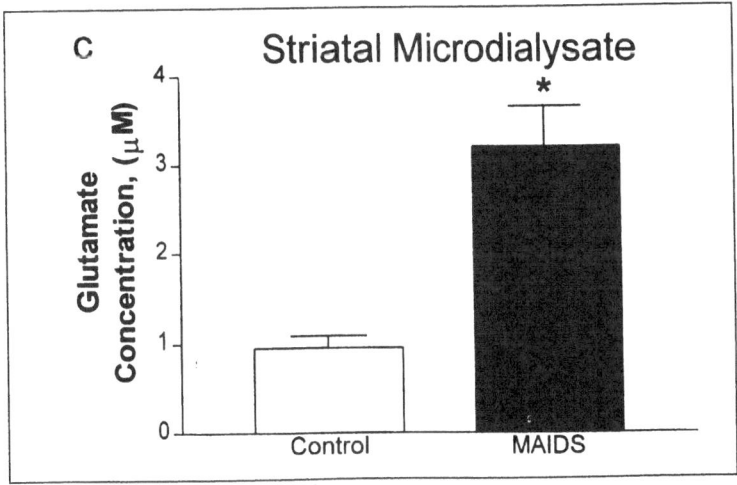

Taken together, the evidence strongly suggests that glutamate itself is the primary excitotoxic agent found in the brains of LP-BM5 infected mice. The free glutamate concentrations in the CNS of infected mice are sufficient to cause neuronal death under acute glucoprivic and hypoxic conditions. Instead, the persistent nature of this neurotoxic insult in the presence of adequate energy reserves results in a chronic neurodegenerative process, leading to neuronal deafferentation, neurotransmitter depletion and compensatory changes in glutamatergic (AMPA) receptors and transporter expression. While this neurodegeneration may be sufficient to account for the cognitive deficits observed in infected mice, additional insults provided by infiltrating leukocytes may be necessary to cause the neuronal death and demyelination leading to the motor abnormalities observed in patients with ADC. Such conditions may occur in mouse strains that are more resistant to the LP-BM5 virus and do not develop as profound a lymphadenopathy as the B6 mice.[12]

Conclusions

Infecting mice with the LP-BM5 retrovirus causes a progressive and ultimately lethal syndrome of immunodeficiency. Associated with this immunodeficiency is the development of an encephalopathy leading to cognitive deficits through indirect (i.e., nonvirally mediated) pathogenic mechanisms. A number of potential neurotoxins are elaborated in, or gain access to, the CNS of LP-BM5 infected mice, but the nature of the histological and biochemical abnormalities observed in these mice is consistent with a chronic, excitotoxic insult. Moreover, the primary excitotoxin appears to be glutamate itself, which may originate from a variety of sources, including activated microglia. Glutamate levels were not correlated with the degree of splenomegaly, an index of disease severity in the periphery. While this observation is consistent with the importance of intrinsic CNS factors to the development of neuronal degeneration, the observation that anti-retroviral drugs can ameliorate the behavioral abnormalities suggests that infection of the peripheral immune system is essential but insufficient for the development of neuropathology. Moreover, the event(s) triggering the activation of microglia and astrocytes, and the source of the excitotoxins, remain unclear.

The ultimate impetus for investigating the neurological abnormalities developed in association with LP-BM5 infection is to understand the mechanisms contributing to the development of ADC, and then create therapeutic modalities for treating this syndrome.

The LP-BM5 infected mouse may be a suitable model, particularly of the early stages of ADC (Table 5.1). Many of the CNS abnormalities observed in the brains of demented HIV-1 patients are found in LP-BM5 infected mice, and their evolution can be readily observed. Moreover, use of these mice allows for the type of invasive biochemical analysis (e.g., microdialysis) that cannot be performed in patients. This has yielded a great deal of information regarding the pathogenesis and the involvement of substances (e.g., PAF, arachidonic acid) that are otherwise too labile or are masked by metabolic roles (e.g., glutamate) to be addressed by analysis of autopsy specimens. In addition, the duration of LP-BM5 infection in B6 mice allows the investigation not only of the indirect mechanisms of neurodegeneration, but also the compensatory changes. Finally, because the neuropathology in LP-BM5 infected B6 mice appears to be reversible (lack of neuronal death), this model is well suited for testing therapeutic interventions. Thus, infection of different mouse strains with the LP-BM5 MuLV mixture may yield valuable insights into the mechanisms contributing to the neuropathology of ADC, as well as chronic neurodegenerative syndromes.

References

1. Mosier DE, Yetter RA, Morse HC III. Retroviral induction of acute lymphoproliferative disease and profound immunosuppression in adult C57BL/6 mice. J Exp Med 1985; 161:766-784.
2. Jolicoeur P. Murine acquired immunodeficiency syndrome (MAIDS): An animal model to study the AIDS pathogenesis. FASEB J 1991; 5:2398-2405.
3. Morse HC III, Chattopadhyay SK, Makino M et al. Retrovirus-induced immunodeficiency in the mouse: MAIDS as a model for AIDS. AIDS 1992; 6:607-621.
4. Laterjet R, Duplan JF. Experiments and discussion on leukemogenesis by cell-free extracts of radiation-induced leukemia in mice. Int J Radiat Biol Relat Stud Phys Chem Med 1962; 5:339-344.
5. Legrand E, Dalculsi R, Duplan JF. Characteristics of the cell populations involved in extrathymic lymphosarcoma induced in C57Bl/6 mice by RadLV-Rs. Leuk Res 1981; 5:223-233.
6. Chattopadhyay SK, Morse HC III, Makino M et al. Defective virus is associated with induction of murine retrovirus-induced immunodeficiency syndrome. Proc Natl Acad Sci USA 1989; 86:3862-3866.
7. Chattopadhyay SK, Sengupta DN, Fredrickson TN et al. Characteristics and contributions of defective, ecotropic, and mink cell focus-inducing viruses involved in a retrovirus-induced immunodeficiency syndrome of mice. J Virol 1991; 65:4232-4241.

8. Huang M, Simard C, Jolicoeur P. Immunodeficiency and clonal growth of target cells induced by helper-free defective retrovirus. Science 1989; 246:1614-1617.

9. Huang M, Jolicoeur P. Characterization of the *gag*/fusion protein encoded by the defective Duplan retrovirus inducing murine acquired immunodeficiency syndrome. J Virol 1990; 64:5764-5772.

10. Dupraz P, Rebal N, Klein SJ et al. The murine AIDS virus gag precursor protein binds to the SH_3 domain of c-abl. J Virol 1997; 71:2615-2620.

11. Hartley JW, Fredrickson TN, Yetter RA et al. Retrovirus-induced murine acquired immunodeficiency syndrome: Natural history of infection and differing susceptibility of inbred mouse strains. J Virol 1989; 63:1223-1231.

12. Huang M, Simard C, Kay DG et al. The majority of cells infected with the defective murine AIDS virus belong to the B-cell lineage. J Virol 1991; 6562-6571.

13. Huang M, Simard C, Jolicoeur P. Susceptibility of inbred strains of mice to murine AIDS (MAIDS) correlates with target cell expansion and high expression of defective MAIDS virus. J Virol 1992; 66:2398-2406.

14. Morse HC III, Giese N, Morawetz R et al. Cells and cytokines in the pathogenesis of MAIDS, a retrovirus-induced immunodeficiency syndrome of mice. Springer Semin. Immunopathol 1995; 17:231-245.

15. Klinken HP, Fredrickson TN, Hartley JW et al. Evolution of B cell lineage lymphomas in mice with a retrovirus-induced immunodeficiency syndrome, MAIDS. J Immnunol 1988; 140:1123-1131.

16. Cerny A, Hugin AW, Hardy RR et al. B cells are required for induction of T cell abnormalities in a murine retrovirus-induced immunodeficiency syndrome. J Exp Med 1990; 171:315-320.

17. Mosier DE, Yetter RA, Morse HC III. Functional T lymphocytes are required for a murine retrovirus-induced immunodeficiency disease (MAIDS). J Exp Med 1987; 165:1737-1742.

18. Kim WK, Tang Y, Kenny JJ et al. In murine AIDS, B cells are early targets of defective virus and are required for efficient infection and expression of defective virus in T cells and macrophages. J Virol 1994; 68:6767-6769.

19. Giese NA, Giese T, Morse HC III. Murine AIDS is an antigen-driven disease: Requirements for major histocompatibility complex class II expression and $CD4^+$ T cells. J Virol 1994; 68:5819-5824.

20. Kanagawa O, Gayama S, Vauprel B. Functional and phenotypic changes of T cells in murine acquired immune deficiency. J Immunol 1994; 152:4671-4679.

21. Sei Y, Arora PK, Skolnick P et al. Spatial learning impairment in a murine model of AIDS. FASEB J 1992; 6:30008-30013.

22. Stewart CA, Morris RGM. The watermaze. In: Sahgal A, ed. Behavioral Neuroscience: A Practical Approach. Oxford: Oxford University Press, 1993;107-122.

23. Paul IA, Heyes MP, Saito K et al. Zidovudine-AZT-treatment in a murine model of AIDS dementia complex. Effects on spatial learning, quinolinic acid levels and NMDA receptor. In: 32nd Annual Meeting of American College of Neuropsychopharmacology. 1993:139 [Abstract].

24. Sei Y, Makino M, Vitkovic L et al. Central nervous system infection in an murine retrovirus-induced immunodeficiency syndrome. J Neuroimmunol 1992; 37:131-140.

25. Sei Y, Kustova Y, Li Y et al. The encephalopathy associated with murine acquired immunodeficiency syndrome. Ann NY Acad Sci 1997; In Press.

26. Budka H, Wiley CA, Kleihues P et al. HIV-Associated disease of the nervous system: Review of the nomenclature and proposal for neuropathology-based terminology. Brain Pathol 1991; 1:143-152.

27. Harouse JM, Wrobleswsha Z, Laughlin MA et al. Human choroid plexus cells can be latently infected with human immunodeficiency virus. Ann Neurol 1989; 25:406-11

28. Falangola MF, Hanly A, Galvao-Castro B et al. HIV infection of human choroid plexus: A possible mechanism of viral entry into the CNS. J Neuropathol Exp Neurol 1995; 54:497-503.

29. Espey MG, Tang Y, Morse HC III et al. Localization of quinolinic acid in the MAIDS model of retrovirus-induced immunodeficiency: Implications for neurotoxicity and dendritic cell pathogenesis. AIDS 1996; 10:151-158.

30. Simard C, Huang M, Jolicoeur P. Murine AIDS is initiated in the lymph nodes draining the site of inoculation, and the infected B cells influence T cells located at distance, in noninfected organs. J Virol 1994; 68:1903-1912.

31. Kustova Y, Sei Y, Goping G et al. Gliosis in the LP-BM5 murine leukemia virus-infected mouse: An animal model of retrovirus-induced dementia. Brain Res 1997; 742:271-282.

32. Eddleston M, Mucke L. Molecular profile of reactive astrocytes-implications for their role in neurologic disease. Neurosci 1993; 54:15-36.

33. Selmaj K, Shafit-Zagardo B, Aquino DA et al. Tumor necrosis factor-induced proliferation of astrocytes from mature brain is associated with down-regulation of glial fibrillary acidic protein mRNA. J Neurochem 1991; 57:823-830.

34. Piani D, Frei K, Do K-Q et al. Murine brain macrophages induce NMDA receptor mediated neurotoxicity in vitro by secreting glutamate. Neurosci Lett 1991; 133:159-162.

35. Ha J-H, Sei Y, Basile AS. Striatal met-enkephalin and substance-P levels are decreased in mice infected with the LP-BM5 murine leukemia virus. J Neurochem 1995; 64:1896-1898.

36. Li Y, Kustova Y, Sei Y et al. Regional changes in constitutive, but not inducible NOS expression in the brains of mice infected with the LP-BM5 leukemia virus. Brain Res 1997; 752:107-116.

37. Everall IP, Hudson L, Al-Sarra S et al. Decreased expression of AMPA receptor messenger RNA and protein in AIDS: A model for HIV-associated neurotoxicity. Nature Med 1995; 1:1174-1178.

38. Kustova Y, Espey, MG, Sei Y et al. Regional decreases in AMPA receptor density in mice infected with the LP-BM5 murine leukemia virus. Neuroreport 1997; 8:1243-1247.

39. Sei Y, Whitesell L, Kustova Y et al. Altered hippocampal *fyn*-kinase in a retroviral-induced immunodeficiency syndrome. FASEB J 1996; 10:339-344.

40. Grant SGN, O'Dell TJ, Karl KA et al. Impaired long-term potentiation, spatial learning, and hippocampal development in fyn mutant mice. Science 1992; 258:1903-1910.

41. Masliah E, Ge N, Morey M et al. Cortical dendritic pathology in HIV encephalitis. Lab. Invest 1992; 66:285-291.

42. Nagra RM, Masliah E, Wiley CA. Synaptic and dendritic pathology in murine retroviral encephalitis. Exp Neurol 1993; 124:283-288.

43. Bukrinsky MI, Nottet HSLM, Schmidtmayerova H et al. Regulation of nitric oxide synthase activity in human immunodeficiency virus type 1 (HIV-1)-infected monocytes: Implications for HIV-associated neurological disease. J Exp Med 1995; 181:735-745.

44. Gelbard HA, Nottet HSLM, Swindells S et al. Platelet-activating factor: A candidate human immunodeficiency virus type-1 induced neurotoxin. J Virol 1994; 68:4628-4635.

45. Nishida K, Markey SP, Skolnick P et al. Increases in brain platelet activating factor were inhibited by the N-methyl-D-aspartate antagonist MK-801 in mice infected with the LP-BM5 murine leukemia virus. J Neurochem 1996; 66:433-435.

47. Stone TW. Neuropharmacology of quinolinic and kynurenic acids. Pharmacol Rev 1993; 45:309-379.

48. Sei Y, Paul IA, Saito K et al. Quinolinic acid levels in a murine retrovirus-induced immunodeficiency syndrome. J Neurochem 1996; 66:296-302.

49. Patneau DK, Mayer ML. Structure-activity relationships for amino acid transmitter candidates acting at N-Methyl-D-aspartate and quisqualate receptors. J Neurosci 1990; 10:2385-2399.

50. Castillo J, Davalos A, Naveiro J et al. Neuroexcitatory amino acids and their relation to infarct size and neurological deficit in ischemic stroke. Stroke 1996; 27:1060-1065.

51. Uchiyama-Tsuyuki Y, Araki H, Yae T et al. Changes in the extracellular concentrations of amino acids in the rat striatum during transient focal cerebral ischemia. J Neurochem 1994; 62:1074-1078.

52. Volterra A, Trotti D, Cassutti C et al. High sensitivity of glutamate uptake to extracellular free arachidonic acid levels in rat cortical synaptosomes and astrocytes. J Neurochem 1992; 59:600-606.

CHAPTER 6

Transgenic Mice Expressing Cytokines in the CNS as Model Systems for the Study of Inflammatory Neurodegenerative and Demyelinating Disorders

Axel Pagenstecher, Eliezer Masliah, Anna K. Stalder,
Iain L. Campbell

Introduction

Cytokines are a diverse group of small molecular weight, soluble factors that were originally discovered as products of activated immune cells and which play a central role in immune regulation.[1,2] However, it is now clear that these multifunctional factors are produced by and act on not only immune cells but virtually all cell types examined. Like hormones, cytokines exert their actions by binding to specific receptors on the surface of cells activating intracellular signaling pathways that result in the modulation of gene transcription. Unlike hormones which are largely endocrine regulators, cytokines tend to act in a more localized milieu influencing the function of immediately neighboring cells (termed paracrine regulation) or the producer cells themselves (termed autocrine regulation). In vivo, cytokines exert their actions within the context of a highly complex and yet tightly regulated network in which the cellular response evoked represents the sum of the overlapping, synergistic and antagonistic actions of multiple cytokines.

Neuroimmunodegeneration, edited by Paul K.Y. Wong and William S. Lynn.
© 1998 Springer-Verlag and R.G. Landes Company.

There is a growing body of evidence that, as in the periphery, cytokines are also multifunctional effector molecules in the CNS.[3-5] Production of cytokines within the CNS may emanate from many sources including infiltrating mononuclear cells such as T cells and macrophages. Significantly, these factors may also be produced by neural cells, which thus provide a local source for cytokine production in the CNS. In particular, astrocytes and microglia are prodigious cytokine producing cells which, in addition to their ability to produce cytokines, also respond to these very same mediators, possibly establishing in the CNS regulatory loops analogous to those found in the immune system. Local CNS production of many proinflammatory cytokines such as IL-1α/β, IL-6, IL-12 and TNFα, as well as counterinflammatory (e.g., IL-10 and TGFβ) and hematopoietic cytokines such as M-CSF and GM-CSF, has been demonstrated and suggest the existence of a neural cytokine network. Considering the multitude of effects that cytokines exert, it is not surprising that the expression of these molecules in the CNS, as in other organs, is normally tightly regulated and maintained at low levels. However, dysregulation of cytokine production can occur in a number of CNS disorders. Expression of various cytokine genes in the brain is known to be altered in a variety of significant neurological disorders such as multiple sclerosis, subacute sclerosing panencephalitis, NeuroAIDS, stroke, Alzheimer's disease, subacute spongiform encephalopathies, and infectious diseases. In many cases overlapping expression of cytokines with areas of neuropathology has been demonstrated. Yet, it is still unclear what role cytokines play in the etiopathology of these neuropathologic states, in particular, whether expression of cytokines in the CNS is pathogenic or protective in these disorders.

The idea that cytokines may be causal agents in CNS disease has gained strong indirect support from a large number of experimental studies on the effects of various cytokines following either their addition to neural cell cultures or their direct injection into the CNS. Unfortunately, these types of experimental approaches while having been informative have a number of drawbacks that could lead to a questioning of the relevance of the findings. For example, in vitro neural cell preparations lack the highly complex cellular organization and interactions of the intact CNS. While infusion of cytokines into the CNS overcomes to some extent the problems of in vitro preparations, this approach results in traumatic injury to the brain, and the nature of cytokine delivery (i.e., a high concentration released into the extracellular space for a short time period) is relatively unphysiologic. However, the development of experimental procedures that permit the stable germline transmission in mice (so-called transgenic mice) of specific genes with expression targeted to the intact CNS, offers a powerful and more relevant approach for studying the CNS actions of cytokines.

Importantly, the transgenic approach permits a specific cytokine gene of interest to be expressed in a milieu in which the complex anatomic and physiological interactions of the CNS are maintained. Once made, stable expressing lines of transgenic mice provide a virtually unrestricted source of identical animals, thus permitting systematic multi-level analysis of pathological, electrophysiological, neuroendocrinological and behavioral manifestations. Direct evidence for the neuropathogenicity of cytokines has now come from the application of transgenic technology to specifically target the expression of cytokine genes to the CNS of mice (for a review see ref. 6). In this chapter, we will briefly discuss the methods used for the generation of transgenic mice. We will then summarize the neuropathological features and behavioral alterations in transgenic mice with CNS targeted expression of the different cytokines.

Generation of Transgenic Mice

There are a variety of methods for introducing new genes or modifying endogenous gene expression in mutant animals. In general, the aim of germline manipulation is the generation of mutant mice with gene expression either added (here referred to as transgenic mice) or removed ('knock out' mice). In recent years, elegant modifications of these basic principles have been developed, allowing for the 'switching' on or off of particular genes at defined times in the life of the animal.[7-8] Here we will focus on the organ or cell type specific, gene promoter-directed (or transgenic) expression of genes.

To develop transgenic mice, a hybrid DNA (or fusion gene) construct is engineered that, when properly integrated into the eukaryotic chromosome, directs the cell specific expression of an introduced (termed transgene) gene. For an appropriate expression of the transgene it is necessary that the construct contain elements for the correct transcriptional, posttranscriptional and translational processing by the targeted cell. Thus, the transgene construct generally contains an upstream transcriptional control unit (the promoter connected to the coding sequence (the expressed gene) and a downstream regulatory sequence (the polyadenylation signal) to facilitate correct processing of the primary RNA transcript. Further modifications of this basic fusion gene cassette, such as the inclusion of splice signaling sequences, can be made and may improve the efficiency of transgene expression. By choosing an appropriate promoter it is possible to direct the transcription of the transgene to particular cell types of interest. Expression under the control of some promoters, e.g., the metallothionein promoter[9] includes but is not restricted to the CNS, while others, e.g., the GFAP promoter (which directs expression to astrocytes)[10] or the NSE promoter (which directs expression to neurons),[11] are largely specific to the CNS. The decision as to which promoter to use (and therefore which neural cell to target for transgene expression) depends to a large extent on the experimental objective and the nature of the transgene product to be expressed. In our case we wished to model, as closely as possible, the inflammatory state of the CNS where chronic expression of cytokines-occurs. As indicated above, both microglia and astrocytes are relevant in this context, since these cells are known to be participant in host responses in the CNS and are a major local source for cytokine production.[5-12] This latter point is also of importance since it means that these glial cells are equipped with the necessary machinery for the synthesis and secretion of the biologically active cytokine. At present, due to the close relationship of microglia and macrophages, there are no gene promoter constructs that direct the transcriptional targeting of microglia, to the exclusion of macrophages. We therefore have made use of fusion gene constructs in which the DNA coding region of the cytokine gene was placed under the transcriptional control of the GFAP promoter, thereby targeting cytokine gene expression to astrocytes. A typical GFAP-cytokine fusion gene construct is depicted in Figure 6.1.

After construction and purification, the transgene is microinjected into the pronucleus of fertilized eggs. These are then implanted in the oviduct of pseudopregnant foster mothers. Offspring are screened for transgene integration, mated and the resultant offspring analyzed for transgene expression. Subsequently, positive mice are further bred to evolve lines of mutant mice with stable integration of the transgene. The integration of the injected transgene DNA into the mouse genome is a random event that can have a number of important consequences. The integration site can influence the level

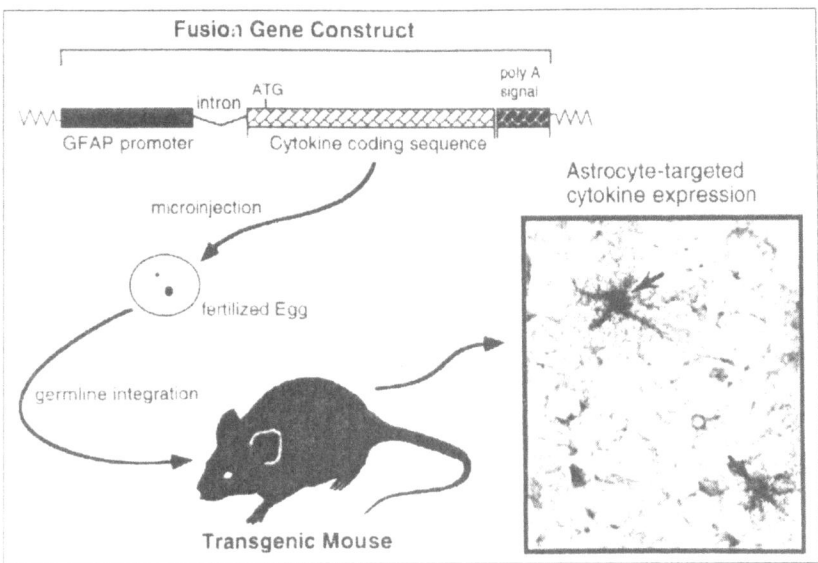

Fig. 6.1. Transgene-targeted expression of cytokines by astrocytes in the mouse central nervous system. In the authors' studies, transcriptional regulation of cytokine gene expression is achieved with the use of fusion gene constructs in which a coding sequence of the cytokine gene is placed downstream of the murine GFAP promoter and upstream of a polyadenylation (poly A) signal sequence. The fusion gene construct is then microinjected into the pronucleus of fertilized eggs, which are subsequently implanted in the oviduct of pseudopregnant recipient mice. Progeny are analyzed for transgene integration and positive mice subsequently mated, their offspring screened for transgene expression and, where positive, used to develop transgenic lines with stable integration and expression of the transgene. In the example provided here, in situ hybridization for TNFα RNA was combined with immunostaining for GFAP to demonstrate astrocyte expression (arrows) of this cytokine in the brain of GFAP-TNFα transgenic mice.

of transgene expression (with variations from zero to extremely high) which can give rise to several independent transgenic lines with different expression levels of the same transgene. This can have several implications. First, it permits the establishment of transgene dose-effect relationships. Second, confirmation can be attained that the expression of the transgene is responsible for the observed alterations in transgenic animals, and that they are not due to some other mechanism such as integration site-induced mutations or disruption of the transcriptional activity of an endogenous gene. Third, in the case of low expression, this often permits the derivation of transgenic lines for highly active or biologically potent molecules (e.g., cytokines) where high expression levels would normally be expected to be toxic.

Transgenic Expression of Cytokines in the CNS

Using the strategies outlined above, the expression of a number of cytokines including IL-6,[13] IL-3,[14] TNFα,[15-17] IFNα,[18] IFNγ,[19] IL-12 (Pagenstecher and Campbell, unpublished), TGFb[20-21] and the chemoattractant cytokines MCP-1[22] and N51/KC[23] have been successfully targeted to the CNS under the transcriptional control of various promoters active in different neural cells (see Table 6.1).

IL-6

IL-6 is a pluripotent cytokine that is involved in the regulation of inflammation, hematopoiesis and acute-phase responses.[24-25] IL-6 is known to be produced by a variety of cell types, including fibroblasts, endothelial cells, macrophages and epithelial cells. Importantly, within the CNS both microglia and astocytes have been demonstrated to be sources for the neural production of this cytokine. Expression of IL-6 is increased in a number of neurological diseases, including neuroinflammatory disorders such as multiple sclerosis and viral and bacterial meningitis, as well as neurodegenerative disorders such as Alzheimer's disease, stroke and NeuroAIDS. Although IL-6 may exert a variety of actions in the CNS, from effects on the differentiation of specific cells such astrocytes and neurons to the coordination of inflammatory responses and modulation of physiologic pathways, e.g., hypothalamic-pituitary-adrenal axis and thermoregulation, the role of cerebrally expressed IL-6 in the pathogenesis of these neurological disorders is not clear. Significant insights into this problem have come from studies in transgenic mice (termed GFAP-IL6 mice) which serve as a model for the utility of this experimental approach, and which will be discussed further here.

We used a GFAP genomic DNA construct to target expression of the cytokine IL-6 to astrocytes in the CNS.[13] Transgene encoded IL-6 RNA expression in the brain of several high expressor founders, and in transgenic offspring from two independent lines that were established from low expressor founder animals, was predominantly in subcortical areas such as the thalamus, the cerebellum and the brain stem. The highest levels of transgene expression were found in the Bergmann glia of the cerebellum. Production of bioactive IL-6 protein was demonstrated in supernatants derived from primary astrocyte cultures isolated from the GFAP-IL6 mice. Importantly, the amount of IL-6 RNA present in the brain of GFAP-IL6 mice was

Table 6.1. **Transgenic models for the study of cytokine actions in the CNS**

Cytokine	Promoter	Targeted Cell	Phenotype	
			Pathological	Clinical
IL-3	GFAP	astrocyte	macrophage/microglial mediated-demyelination	adult (>5 mo) onset progressive motor disorder/premature death
IL-6	GFAP	astrocyte	local inflammation/ neurodegeneration	adult (>3 mo) onset progressive learning deficit/tremor/seizure
IFN-α	GFAP	astrocyte	chronic encephalitis/ calcification neurodegeneration	progressive wasting/ ataxia/premature death
IFN-γ	MBP	oligoden-drocyte	hypomyelination /disrupted cerebellar development	severe postnatal/ ataxia/premature death
MCP-1	MBP	oligoden-drocyte	perivascular mono-nuclear cell infiltrates	none
N51/KC	MBP	oligoden-drocyte	neutrophil infiltration /microgliosis/BBB breakdown/neurological symptoms	adult onset (>6 mo) ataxia/wasting/ premature death
TGF-β	GFAP	astrocyte	hydrocephalus/excessive extracellular matrix formation	postnatal runting/ ataxia/premature death
TNF-α	TNF-α	neuron	encephalomyelitis/ demyelination	postnatal wasting/ paralysis/premature death
	GFAP	astrocyte	lymphocytic encephalomyelitis/ demyelination/ neurodegeneration	adult onset (>3 mo) paralysis/wasting/ premature death

similar to that found in the brain in EAE.[26] Thus, the level of expression of the transgene encoded IL-6 in the CNS of the GFAP-IL6 mice is within a pathophysiologic rather than a pharmacologic range.

GFAP-IL6 mice developed a neurologic syndrome in which both the clinical symptoms and histological alterations related to the level and distribution of transgene expression. In a low expressor GFAP-IL6 line, mice show no overt physical abnormalities until around 10-12 mo age, from which time many animals exhibit progressive deterioration in health and also develop minor tremor, with some animals progressing on to develop ataxia and seizures. Behavioral and neurophysiological assessment of these GFAP-IL6 animals demonstrated a positive correlation between functional impairments and transgene dose. Moreover, the symptoms were progressive; for example, deficits in avoidance learning were increased with age of the transgenic mice.[27] Neurophysiological dysfunction is present in these transgenic mice, which have anomalous hippocampal paroxysmal discharges and suppressed theta rhythm.[28] A significantly reduced long-term potentiation in the dentate is a further electrophysiological alteration detected in hippocampal slices from GFAP-IL6 homozygous mice.[29] The disruption of long-term potentiation together with the suppression of theta rhythm in the GFAP-IL6 mice are indicative of disrupted synaptic plasticity and could underlie the altered cognitive function observed.

Detailed histopathological analysis was performed to correlate the behavioral and electrophysiological changes to morphological alterations. In the hippocampus, dendritic vacuolization, stripping of dendritic spines and decreased synaptic density were observed (see Fig. 6.2). Moreover, there was a significant decrease in the number of parvalbumin and calbindin immunoreactive neurons that are a major source of the inhibitory neurotransmitter GABA. Thus, it is likely that these neurodegenerative changes, at least in part, underlie the learning deficits and hippocampal pathophysiology in the GFAP-IL6 mice. Neurodegenerative changes were also present in the cerebellum, where atrophy and loss of molecular and granular layer neurons was pronounced. At the ultrastructural level, spongiosis and pronounced axonal dystrophy with secondary demyelination were observed. These degenerative changes in the cerebellum correlated well with the development of motor abnormalities in some old transgenic animals. Additionally, structural and functional changes to the cerebrovascular endothelium are prevalent. GFAP-IL6 mice

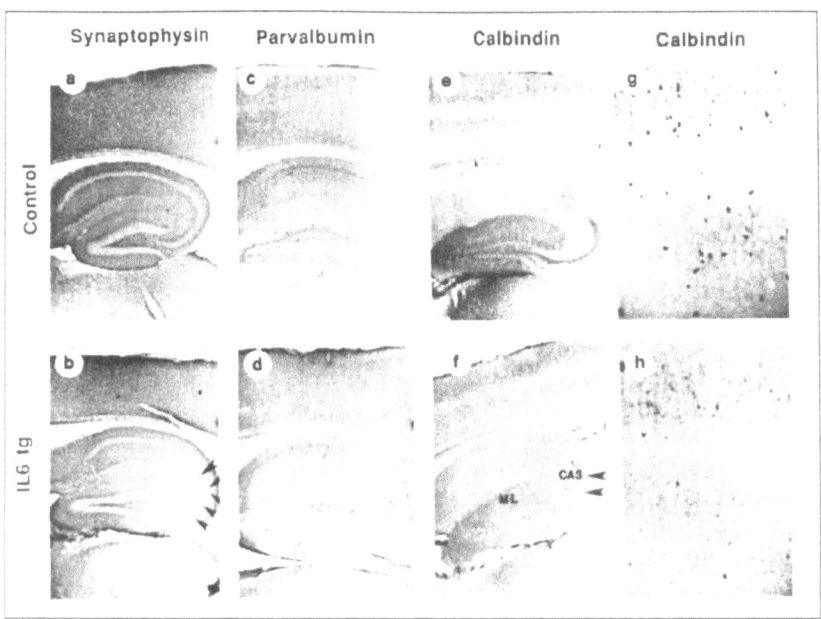

Fig. 6.2. Patterns of neurodegeneration in GFAP-IL6 transgenic mice. In control mice (a) there was a strong immunolabeling of the neuropil with antibodies against synaptophysin; in contrast, in IL6 transgenic mice (b) there was a reduction in the CA2 region of the hippocampus (arrowheads). In addition to the synaptic loss, compared to controls (c,e,g), IL6 transgenic mice also showed widespread reduction in parvalbumin (d) and calbindin immunoreactive neurons in the hippocampus (f) and neocortex (f,h).

do not develop a functional BBB (blood-brain barrier) and have proliferative angiopathy and increased expression of von Willebrand factor in the brain.[30-31] Altogether these pathological alterations may further contribute to neuronal injury. Interestingly, transgenic expression of IL-6 in the CNS provokes only minor responses by mononuclear cells, with mainly B cells accumulating around some larger vessels in the cerebellar sulci and the brain stem. However, a pronounced chronic inflammatory response is evident and involves neural cells in the GFAP-IL6 mice.[13,32-34] This response includes upregulation or induction of mRNAs for the acute-phase response genes α1-antichymotrypsin, complement C3 and metallothionein I + II, together with expression of other cytokine genes including IL-1α and β and TNFα and the increased expression of ICAM-1. Overlapping with areas that show neurodegenerative changes, diffuse astrogliosis and an activation of microglia are also present.

These findings in the GFAP-IL6 transgenic mice indicate that chronic expression of IL-6 in the CNS may result in neuro-degenerative disease with disruption of electrophysiological, neuroendocrine and learning functions. Several of the structural and functional alterations observed in GFAP-IL6 transgenic mice resemble the features seen in human neuroinflammatory disorders such as NeuroAIDS or Alzheimer's disease. Thus GFAP-IL6 transgenic mice could serve as a model for some aspects of human disorders in which a dysregulation of proinflammatory cytokines occur.

TNFα

TNFα is a potent proinflammatory cytokine and is the primary mediator of endotoxic shock.[35-36] TNF production can be stimulated (especially by gram negative bacteria) from a wide variety of cells. The cerebral expression of TNFα is increased in a variety of neurological disorders, including NeuroAIDS, bacterial meningitis, cerebral malaria, cerebral ischemia and multiple sclerosis. Experimentally, TNFα expression in the CNS is closely associated with the development of experimental autoimmune encephalomyelitis (EAE) and cerebral malaria. Furthermore, strategies that block the production or activity of TNFα in these models also produce significant reductions in CNS inflammation and clinical severity. The function of TNFα and its contribution to the pathogenesis of these inflammatory CNS disorders has recently been the subject of studies in transgenic mice.

Transgenic mice have been generated expressing a murine TNFα transgene under the control of its own promoter.[15] These mice exhibit marked neurological impairment and require treatment with a neutralizing anti-murine TNFα antibody to stay alive. Although the TNFα transgene is expressed under the control of its own promoter, transgene expression, curiously, localizes predominantly to CNS neurons. Transgenic mice present with early onset of ataxia, seizures, paresis and premature death. Histologically, CD4[+] and CD8[+] lymphocytic meningitis and encephalitis, diffuse astrocytosis and microgliosis, and demyelination occur. Other degenerative changes (e.g., in neurons) were not reported, although these might be expected given the reported severity of the meningoencephalitis. Although neurons have been reported to express TNFα immunoreactive protein in the normal brain,[37] the expression of this cytokine in neuroinflammatory disorders has been localized largely to astrocytes and microglia and not neurons.[5,38] In order to target CNS TNFα expression to a seemingly more appropriate cell type, transgenic mice have been developed that express human or murine TNFα specifically in astrocytes under the control of the GFAP-promoter.[16-17] The findings in these GFAP-TNF mice are comparable to those in transgenic mice expressing TNFα in neurons under the control of its own promoter as discussed above. Transgenic expression of TNFα in astrocytes results in the development of a neurological disorder with symptoms ranging from ataxia to complete hindlimb paralysis and premature death. The histological hallmark is meningoencephalomyelitis with extensive infiltration of the CNS parenchyma with CD4[+] and CD8[+] lymphocytes. Demyelination as well as neurodegeneration occurs in areas of inflammation. Moreover, the cerebrovascular endothelium is highly activated with upregulation of cellular adhesion molecules such as ICAM-1. Pronounced apoptosis is also associated with infiltrating inflammatory cells, indicating active CNS turnover of these cells.

Altogether, the findings in these transgenic models corroborate the proinflammatory actions of TNFα and demonstrate the ability of this cytokine to promote the recruitment and trafficking of inflammatory cells into the CNS. These transgenic models may serve as valuable tools in our understanding of the pathogenesis of human neurological disorders such as cerebral malaria and MS that are associated with lymphocytic encephalomyelitis and where elevated TNFα gene expression occurs.

The Interferons

The interferons (IFNs) are important host response cytokines with antiviral, antitumor and immunoregulatory functions. Based on their properties and cellular receptors, IFNα form two groups (for reviews see refs. 39-40). Type I IFNs include IFNα and IFN-β, which share a common receptor and have closely overlapping biological actions. At present, >20 IFNα genes have been identified, of which at least 16 encode separate protein products. In contrast, there is only one gene for IFN-β. The type I IFNs are produced by a wide variety of cells, including leukocytes (lymphocytes and macrophages), fibroblasts and epithelial cells. Within the brain, potential cellular sources for localized production of IFNs include astrocytes and microglia. Type II IFN, which is commonly known as IFNγ, shows a more restricted distribution compared with the type I IFNs, and is expressed by activated T lymphocytes and natural killer cells and binds to a unique receptor. Although sharing many overlapping actions with the type I IFNs, IFNγ also possesses several activities which are unique and relate particularly to its immunoregulatory functions, e.g., induction of MHC class II expression. It is now known that in addition to their antiviral actions, IFNs of both types may profoundly affect a variety of functions, including cellular metabolism, growth and differentiation, immune function and tumor development.

In relation to possible involvement in neurological diseases, IFNα therapy in humans is associated with major CNS complications, while intrathecal synthesis of IFNα is linked to the familial progressive encephalopathy Aicardi-Goutières syndrome.[41] To better define the CNS actions of type I interferons and their role in host defense to infection with neurotropic viruses, we recently generated transgenic mice with astrocyte-targeted expression of IFNα1 under the control of the GFAP-promoter.[18] In one line of GIFN-IFNα mice, mice are runted, moribund and die at 5-10 months of age, while a similar, but much less severe, phenotype is observed in some older (8-12 mo) transgenic mice from another line with significantly lower expression of IFNα. In the brain of transgenic mice, a spectrum of alterations is observed that correlate with the levels and distribution of transgene expression. At the cellular level, gliosis, angiopathy with mononuclear cell cuffing and progressive calcification affecting basal ganglia and cerebellum are prominent. Spongiform changes are present in regions affected by calcification. At the molecular level, the expression of two IFNα inducible genes, 2',5'-oligoadenylate synthetase and MHC class I, is markedly increased. These findings demonstrate that the chronic cerebral expression of IFNα induces an inflammatory encephalopathy and causes progressive CNS injury. Many of the cellular alterations found in the CNS of the GFAP-IFNα mice overlap with those reported in the Aicardi-Goutières syndrome, therefore substantiating the suggestion that IFNα is the primary neuropathogenic mediator in this syndrome.[41]

Expression of IFNγ in the CNS occurs in a number of human (e.g., MS) and experimental (e.g., EAE and the viral meningoencephalitides) neurological disorders where mononuclear cell infiltration is a prominent feature. Transgenic mice have been developed in which the expression of IFNγ was placed under the transcriptional control of the MBP gene.[19] MBP-IFNγ transgenic mice show a shaking/shivering phenotype with dramatic hypomyelination and abnormal cerebellar development. Gliosis and increased expression of MHC class I and II in white matter are also observed. Somewhat surprisingly, given the potent proinflammatory actions of IFNγ, mononuclear cell infiltration in the CNS does not occur in the MBP-IFNγ transgenic mice. The basis for the hypomyelinating phenotype is not known, but clearly does not involve an inflammatory primary demyelination process. In many respects the loss of myelin and the physical presentation in these transgenic mice are similar to those observed for MBP-MHC class I transgenic mice[42] and it is speculated that one explanation for the hypomyelination in the MBP-IFNγ transgenic mice is that it results from the overexpression of MHC class I by the oligodendrocytes. Whether this is the case, or whether the hypomyelination results from direct oligodendrocyte effects of IFNγ or from a nonspecific transgene induced effect remains to be determined.

IL-3

IL-3 is produced by activated T lymphocytes and is a potent macrophage-activating factor in the course of T cell dependent immune responses.[43-44] In the CNS, microglia are known to produce and respond to IL-3.[45-46] In the rat brain, the signal transducing β-subunit of the IL-3 receptor shows restricted expression to macrophage/microglial cells.[47] In vitro, IL-3 induces proliferation and multinucleated giant cell formation of microglia. IL-3 may also serve as a trophic factor in the brain, promoting the survival and differentiation of cholinergic neurons.[48] There is a lack of definitive information concerning the role of IL-3 in physiological and pathological processes in the CNS. To examine these questions, we developed transgenic mice in which the expression of IL-3 was targeted to astrocytes as described above.[14]

GFAP-IL3 mice expressing IL-3 at low levels in the brain develop progressive motor impairment (gait abnormality, tremor, ataxia and later quadriplegia) starting between 4 and 5 months of age. Histologically, in the brain of GFAP-IL3 animals, hypertrophy, proliferation and enhanced expression of MHC class II molecules by macrophage/microglia cells are early changes seen at 2-3 months of age—before clinical symptoms occur. Symptomatic transgenic mice have multi-focal "plaque-like" inflammatory white matter lesions in the cerebellum and brain stem. These lesions consist predominantly of macrophages/microglia. Mast cells and, late in lesion development, lymphocytes were also identified in the white matter lesions. Since a motor disorder with similar pathologic features was observed in GFAP-IL3/SCID mice, T and B lymphocytes are apparently not necessary for the development of this demyelinating disease. Ultrastructural features include primary demyelination with accumulation of large numbers of macrophages/microglia within the lesions. Many of these cells contain numerous lipid rich vacuoles and/or novel crystalline 'pole-shaped' inclusions. The lipid-laden macrophages are frequently observed to circumscribe demyelinated axons, while in other cases macrophages were seen in the process of separating the myelin sheath from the axonal surface. The presence of numerous axons ensheathed by a thin, uniform layer of myelin suggests ongoing remyelination, thereby indicating that in the lesions, demyelination and remyelination are concurrent processes. In contrast to the damaged myelin sheath, oligodendrocyte cell bodies appear healthy. Macrophages/microglia within the lesions were activated, as demonstrated by increased expression of the C3 complement receptor and MHC class II (Ia) molecules.

Altogether, these studies show that chronic low level expression of IL-3 in the CNS promotes the recruitment, extravasation, activation and proliferation of macrophage/microglial cells in the white matter. These early events are followed by a progressive T cell independent, macrophage/microglial mediated demyelinating disease leading to motor impairment and premature death. Thus, chronic stimulation of macrophages/microglia alone can be sufficient for the development of an inflammatory demyelinating disorder. Many of the pathologic features in GFAP-IL3 transgenic mice resemble those observed in HIV-leukoencephalopathy and acute MS. Thus, this transgenic mouse could serve as a useful model to study the role of macrophage/microglia in the pathogenesis of human inflammatory demyelinating disease and for the development of therapeutic strategies aimed at preventing this process.

Other Cytokines

The transforming growth factor beta (TGFβ) family consists of at least three isoforms (called TGFβ1, TGFβ2 and TGFβ3) that are encoded by separate genes and participate in a diverse array of biological processes involved in cellular growth and differentiation.[49] The TGFβ cytokines are expressed in the CNS under physiological as well as pathophysiological conditions; however, their role in the CNS in health and disease is not well understood. Increased TGFβ1 levels have been found in the brains and in astrocytes and microglia of patients with neurodegenerative disorders such as NeuroAIDS and Alzheimer's disease. The relevance of TGFβ expression in the CNS has recently been examined in transgenic mice in which transcription of the porcine TGFβ1 gene was controlled by the human[20] or the mouse[21] GFAP promoters. Essentially identical findings were reported in the two separate studies, with transgenic mice having severe communicating hydrocephalus, seizures, motor incoordination and early death. Consistent with its known actions, astrocyte overexpression of TGFβ1 induces a pronounced increase in extracellular matrix protein production that may contribute to the development of hydrocephalus found in the transgenic animals. These findings indicate that the developing CNS is highly sensitive to TGFβ1, which may influence the developmental fate of the neural cells directly, as well as perturb cell migration at the level of altered extracellular matrix formation.

Leukocyte recruitment and infiltration of the central nervous system (CNS) represents a cardinal feature in the pathogenesis of diverse inflammatory neurological disorders such as bacterial and viral meningoencephalitis, multiple sclerosis, HIV encephalopathy, and cerebral malaria. Chemokines are a novel family of chemoattractant cytokines that are important in leukocyte adhesion to the endothelium and emigration into peripheral and CNS tissues during inflammation.[50-51] The role of chemokines in neuroinflammatory disorders has been studied in transgenic mice. The constitutive expression of the murine chemokine MCP-1[22] or the murine chemokine N51/KC[23] was directed to oligodendrocytes using the MBP promoter. Expression of MCP-1 in the brain produced a mononuclear cell infiltration that is discrete, focal and largely perivascular. The intensity of this infiltrate was increased as well as some parenchymal infiltration stimulated, following systemic LPS administration to the MBP-MCP1 transgenic mice. In all, there is no overt evidence of CNS injury in these transgenic mice, indicating that while MCP-1 can provide a signal for the recruitment of mononuclear cells to the CNS, it does not promote an active inflammatory process. In contrast to MCP-1, expression of N51/KC produced a florid neutrophil infiltration into perivascular, meningeal and parenchymal regions of the CNS. This response is consistent with the known property of N51/KC as a predominantly neutrophil chemoattractant cytokine. Interestingly, as with MCP-1, transgenic expression of N51/KC, while effective at promoting CNS recruitment, apparently failed to induce activation of the infiltrating neutrophils which were not degranulated. However,

CNS neutrophilia in these transgenic mice is associated with major neuropathological alterations which include pronounced micro-gliosis and BBB disruption. Despite these marked pathologic changes, no significant damage to neurons or myelin occurs. Interestingly though, at a later age, well after the peak of chemokine expression, MBP-N51/KC mice develop a neurological syndrome with postural instability and rigidity and premature death. In the absence of sig-nificant neurodegenerative or white matter disease, it is speculated that chronic exposure of neurons to the chemokine may directly compromise their function. Receptors for the related human chemokine IL-8 are known to be expressed on some classes of neu-rons in the CNS, supporting the possibility that this class of chemokines may indeed participate in receptor-mediated modula-tion of neuronal activity.

Concluding Perspectives

In general, the results from the study of these different transgenic models make a number of important points. First, transgenic expression of different cytokines in the CNS, even those sharing some overlapping functions, e.g., IL-6 and TNFα, often pro-duces striking and unique pathological and clinical phenotypes (see Table 6.1). Thus, these findings indicate convincingly that cytokines can act as potent pathogenic agents in the CNS, with the chronic expression of these host-response molecules in the brain inducing an array of neuropathological changes. Moreover, these CNS-cytokine transgenic mice represent powerful tools for unraveling the mechanistic basis for the effects of individual cytokines in the CNS. An important consideration to note at this point is that the timing for transgene encoded cytokine expression in the CNS cannot be manipulated experimentally. Therefore, in most instances, expres-sion of the transgene encoded cytokine is present in and may influ-ence the developing CNS. A good example here is the GFAP-driven expression of TGFβ discussed above, which produces obvious and profound developmental abnormalities. As a solution to this poten-tial complication, recent technological advances in this area offer the prospect of obtaining tight temporal control of gene expression in transgenic mice with the use of antibiotic or hormone responsive binary promoter constructs. These permit the switching of transgene expression on and off simply by administration of the promoter sen-sitive antibiotic or hormone factor to the animal. Second, in many instances the pathological and clinical phenotypes in the transgenic models are similar to those found in human neuroinflammatory disorders. These animal models therefore provide clearcut evidence that not only can cytokines be considered as pathogenic factors in human neuroinflammatory disorders but also they are relevant to our understanding of the role of cytokines in these disease states. Clearly, these transgenic models offer great potential for the identification and development of therapeutic approaches to abrogate the harmful actions of these cytokines in the CNS.

References

1. Balkwill F. Cytokines in health and disease. Immunol Today 1993; 14:149-150.
2. Harrison LC, Campbell IL. Cytokines: An expanding network of immune hormones. Mol Endocrinol 1988; 2:1151-1156.
3. Goetzl EJ, Adelman DC, Sreedharan SP. Neuroimmunology. Adv Immunol 1990; 48:161-190.
4. Plata-Salaman CR. Immunoregulators in the nervous system. Biobehav Rev 1991; 15:185-215.
5. Merrill JE, Benveniste EN. Cytokines in inflammatory brain lesions: Helpful and harmful. Trends Neurosci 1996; 19:331-338.
6. Campbell IL. Neuropathogenic actions of cytokines assessed in transgenic mice. Int J Dev Neurosci 1995; 13:275-284.
7. Furth PA, St Onge L, Boger H et al. Temporal control of gene expression in transgenic mice by a tetracycline-responsive promoter. Proc Natl Acad Sci USA 1994; 91:9302-9306.
8. Kuhn R, Schwenk F, Aguet M et al. Inducible gene targeting in mice. Science 1995; 269:1427-1429.
9. Palmiter RD, Brinster RL. Germ-line transformation of mice. Ann Rev Genet 1986; 20:465-499.
10. Mucke L, Oldstone MBA, Morris JC et al. Rapid activation of astrocyte-specific expression of GFAP-lacZ transgene by focal injury. New Biol 1991; 3:465-474.
11. Forss-Petter s, Danielson PE, Catsicus S et al. Transgenic mice expressing beta-galactosidase in mature neurons under neuron-specific enolase promoter control. Neuron 1990; 5:187-197.
12. Benveniste EN. Inflammatory cytokines within the central nervous system: Sources, function, and mechanism of action. Am J Physiol 1992; 263:C1-C16.
13. Campbell IL, Abraham CR, Masliah E et al. Neurologic disease induced in transgenic mice by the cerebral overexpression of interleukin 6. Proc Natl Acad Sci USA 1993; 90:10061-10065.
14. Chiang C-S, Powell HC, Gold L et al. Macrophage/microglial-mediated primary demyelination and motor disease induced by the central nervous system production of interleukin-3 in transgenic mice. J Clin Invest 1996; 97:1512-1524.

15. Probert L, Akassoglou K, Pasparakis M et al. Spontaneous inflammatory demyelinating disease in transgenic mice showing central nervous system-specific expression of tumor necrosis factor α. Proc Natl Acad Sci USA 1995; 92:11294-11298.

16. Stalder AK, Pagenstecher A, Campbell IL. Lymphocytic meningoencephalomyelitis induced by the transgenic expression of TNFa in the CNS. Soc Neurosci Abstr 1996; 22:1455.

17. Akassoglou K, Probert L, Kontogeorgos G et al. Astrocyte-specific but not neuron-specific transmembrane TNF triggers inflammation and degeneration in the central nervous system of transgenic mice. J Immunol 1997; 158:438-445.

18. Akwa Y, Eloranta ML, Sandberg K et al. Encephalopathy induced by the astrocyte-targeted expression of interferon-α in transgenic mice. Soc Neurosci Abs 1996; 22:278.

19. Corbin JG, Kelly D, Rath EM et al. Targeted CNS expression of interferon-γ in transgenic mice leads to hypomyelination, reactive gliosis, and abnormal cerebellar development. Mol Cell Neurosci 1996; 7:354-370.

20. Galbreath E, Kim S-J, Park K et al. Overexpression of TGFb1 in the central nervous system of transgenic mice results in hydrocephalus. J Neuropathol Exp Neurol 1995; 54:339-349.

21. Wyss-Coray T, Feng L, Masliah E et al. Increased central nervous system production of extracellular matrix components and development of hydrocephalus in transgenic mice overexpressing transforming growth factor-β1. Am J Pathol 1995; 147:53-67.

22. Fuentes ME, Durham SK, Swerdel MR et al. Controlled recruitment of monocytes and macrophages to specific organs through transgenic expression of monocyte chemoattractant protein-1. J Immunol 1995; 155:5769-5776.

23. Tani M, Fuentes ME, Peterson JW et al. Neutrophil infiltration, glial reaction, and neurological disease in transgenic mice expressing the chemokine N51/KC in oligodendrocytes. J Clin Invest 1996; 98: 529-539.

24. Hirano T, Akira S, Taga T et al. Biological and clinical aspects of interleukin 6. Immunol Today 1990; 11:443-449.

25. Le J, Vilcek J. Biology of disease—Interleukin 6: A multifunctional cytokine regulating immune reactions and the acute phase protein response. Lab Invest 1989; 61:588-602.

26. Campbell IL. Structural and functional impact of the transgenic expression of cytokines in the CNS. Ann NY Acad Sci 1997; in press.

27. Heyser CJ, Masliah E, Samimi A et al. Progressive decline in avoidance learning paralleled by inflammatory neurodegeneration in transgenic mice expressing interleukin-6 in the brain. Proc Natl Acad Sci USA 1997; 94:1500-1505.

28. Steffensen SC, Campbell IL, Henriksen SJ. Site-specific hippocampal pathophysiology due to cerebral overexpression of interleukin-6 in transgenic mice. Brain Res 1994; 652:149-153.

29. Bellinger FP, Madamba SG, Campbell IL et al. Reduced long-term potentiation in the dentate gyrus of transgenic mice with cerebral overexpression of interleukin-6. Neurosci Letts 1995; 198:95-98.
30. Brett FM, Mizisin AP, Powell HC et al. Evolution of neuropathologic abnormalities associated with blood-brain barrier breakdown in transgenic mice expressing interleukin-6 in astrocytes. J Neuropathol Exp Neurol 1995; 54:766-775.
31. Campbell IL. Proliferative angiopathy in the brain induced by the cerebral overexpression of interleukin-6 in transgenic mice. J Cell Biochem 1994; 18A(Supplement):283.
32. Chiang C-S, Stalder A, Samimi A et al. Reactive gliosis as a consequence of interleukin-6 expression in the brain. Studies in transgenic mice. Dev Neurosci 1994; 16:212-221.
33. Barnum SR, Jones JL, Muller-Ladner U et al. Chronic complement C3 gene expression in the CNS of transgenic mice with astrocyte-targeted interleukin-6 expression. Glia 1996; 18:107-117.
34. Hernandez J, Molinero A, Campbell IL et al. Transgenic expression of interleukin-6 in the central nervous system regulates brain metallothionein-I and -III expression in mice. Brain Res 1996; 48:125-131.
35. Fiers W. Tumor necrosis factor: Characterization at the molecular, cellular and in vivo level. FEBS Letts 1991; 285:199-212.
36. Le J, Vilcek J. Biology of disease: Tumor necrosis factor and interleukin 1: Cytokines with multiple overlapping biological activities. Lab Invest 1987; 56:234-248.
37. Breder CD, Tsujimoto M, Terano Y et al. Distribution and characterization of tumor necrosis factor-α-like immunoreactivity in the murine central nervous system. J Comp Neurol 1993; 337:543-567.
38. Feuerstein GZ, Liu T, Barone FC. Cytokines, inflammation, and brain injury: Role of tumor necrosis factor-α. Cerebrovas Brain Metabol Rev 1994; 6:341-360.
39. Baron S, Tyring SK, Fleischmann WR et al. The interferons: Mechanisms of action and clinical applications. JAMA 1991; 266:1375-1382.
40. Gutterman JU. Cytokine therapeutics: Lessons from interferon α. Proc Natl Acad Sci USA 1994; 91:1198-1205.
41. Lebon P, Badoual J, Ponsot G et al. Intrathecal synthesis of interferon-alpha in infants with progressive familial encephalopathy. J Neurol Sci 1988; 84:201-208.
42. Turnley AM, Morahan G, Okano H et al. Dysmyelination in transgenic mice resulting from expression of class I histocompatibility molecules in oligodendrocytes. Nature 1991; 353:566-569.
43. Frendl G, Beller DI. Regulation of macrophage activation by IL-3. J Immunol 1990; 144:3392-3399.
44. Frendl G. Interleukin 3: From colony-stimulating factor to pluripotent immunoregulatory cytokine. J Immunol 1992; 14:421-430.
45. Lee TT, Martin FC, Merrill JE. Lymphokine induction of rat microglia multinucleated giant cell formation. GLIA 1993; 8:51-61.

46. Gebicke-Haerter PJ, Appel K, Taylor GD et al. Rat microglial interleukin-3. J Neuroimmunol 1994; 50:203-214.
47. Appel K, Buttini M, Sauter A et al. Cloning of rat interleukin-3 β-subunit from cultured microglia and its RNA expression in vivo. J Neurosci 1995; 15:5800-5809.
48. Kamegai M, Niijima K, Kunishita T et al. Interleukin 3 as a trophic factor for central cholinergic neurons in vitro and in vivo. Neuron 1990; 2:429-436.
49. Wahl SM, McCartney-Francis N, Mergenhagen SE. Inflammatory and immunomodulatory roles of TGFb. Immunol Today 1989; 10:258-261.
50. Schall TJ, Bacon KB. Chemokines, leukocyte trafficking, and inflammation. Curr Opin Immunol 1994; 6:865-873.
51. Ransohoff RM, Glabinski A, Tani M. Chemokines in immune-mediated inflammation of the central nervous system. Cytokine Growth Factor Rev 1996; 7:35-46.

CHAPTER 7

Role for TNF in CNS Inflammation, Demyelination and Neurodegeneration Studied in Transgenic Mice

Katerina Akassoglou, George Kassiotis, George Kollias, and Lesley Probert

Tumor necrosis factor (TNF) is a pluripotent cytokine which is believed to play a central role in the pathogenesis of human immune-mediated diseases. In neuroimmunologic diseases TNF is produced locally in the central nervous system (CNS) and in vitro experiments have shown that it is capable of exerting proliferative and/or cytotoxic effects upon isolated CNS cells depending upon the identity of the target cell. Through the recent application of transgenic and gene knockout technology to the study of TNF neurobiology, it has become possible to directly assess the contribution of TNF action to CNS pathology. Our aim in this chapter is to discuss the cellular interactions and molecular pathways through which TNF can induce neuropathology in vivo. Special emphasis is given to the TNF/p55 TNFR-dependent inflammation, demyelination and neurodegeneration that develop when TNF is chronically expressed by resident CNS cells of transgenic mice. TNF transgenic mice, which model the pathology observed in neuroimmunologic diseases, provide important information concerning the etiopathogenesis of such diseases in humans and support the use of therapeutic approaches designed to target TNF/p55 TNFR signaling pathways in their treatment.

Neuroimmunodegeneration, edited by Paul K.Y. Wong and William S. Lynn.
© 1998 Springer-Verlag and R.G. Landes Company.

Introduction

The nervous system and the immune system are intimately related, both structurally and functionally, and display an active bidirectional communication. This is facilitated by the use of common mediators such as neuropeptides and cytokines.[1] The nervous system can modulate immunological and inflammatory responses, for example through neuropeptide receptors that are located on leukocytes, and is able to influence the severity of inflammatory and autoimmune diseases.[2] On the other hand, cytokines and cytokine receptors classically associated with the peripheral immune system are now known to play important roles in normal CNS functioning,[3] and cytokine disbalances are associated with CNS disease.[4] TNF has become recognized as a pleiotropic cytokine involved in the pathogenesis of several CNS diseases, such as multiple sclerosis (MS),[5] Alzheimer's disease,[6] bacterial meningitis,[7] cerebral malaria[8] and AIDS dementia complex.[9] The cell types that contribute to the local production of TNF within the CNS during the pathogenesis of neuronal disease are resident CNS cells, mainly immunocompetent microglia and astrocytes, and infiltrating cells of the immune system such as T cells and macrophages.[10-12]

A line of in vitro and recent in vivo experiments have attempted to establish the effects of TNF upon individual cell types of the CNS and to correlate them with the pathologic manifestations observed in human diseases. Such experiments have shown for instance the cytotoxic effects of TNF on oligodendrocytes and myelin[13,14] and its proliferative effects on astrocytes[15] and microglia.[16] Recent evidence has associated members of the TNF receptor superfamily such as p55 TNFR and Fas with the transduction of cell death signals in oligodendrocytes in MS.[17,18] TNF also induces the expression of adhesion molecules on endothelial cells[19,20] and astrocytes,[21] thereby promoting the trafficking of inflammatory leukocytes into the brain. Moreover, TNF upregulates the expression of major histocompatibility complex class I,[22] class II[23] and costimulatory molecules[24] on astrocytes, thereby endowing them with antigen presenting cell properties.[25,26] Current research has also revealed an intriguing role for TNF in neuron survival. In vitro data show that TNF can exert toxic[27] and protective effects on neurons,[28,29] and a recent in vivo study gives evidence for a neuroprotective role for TNF during excitotoxic and ischemic brain injury.[30]

While these studies indicate the potential importance of TNF action during CNS disease, its contribution to pathology in vivo, and the cellular and molecular mechanisms by which its effects are mediated in the CNS, remain elusive. Major questions are whether the expression of TNF by resident CNS cells in vivo is sufficient to trigger neurologic disease and whether all CNS cells are similarly capable of triggering neuropathologic alterations. Additional questions concern the differential action of the two biologically active molecular forms of TNF, the 26 kDa transmembrane precursor protein[31] which acts locally through cell-to-cell contact,[32] and the secreted 17 kDa form which is cleaved from the transmembrane form by proteolytic enzymes[33,34] and is capable of acting at distant targets. In the periphery these two molecular forms of TNF are known to mediate distinct biological functions. LPS-induced shock, for example, is established as an effect of soluble TNF,[35] while transmembrane TNF is proposed to be the main effector molecule in immunological reactions such as anti-leishmanial defense in macrophages,[36] T cell-B cell interactions,[37] tumor cell killing mediated by infiltrating lymphocytes,[38] and hepatocellular necrosis and apoptosis in hepatitis.[39] The strict compartmentalization of CNS tissues provides a unique opportunity for analyzing the importance of cellular contacts between the TNF-producing cells and the target cells of TNF in the CNS in vivo.

TNF function is further complicated at the receptor level. TNF signaling is mediated by two specific high affinity receptors, the p55 TNF receptor (TNFR) and the p75 TNFR. The two receptors are coexpressed in most tissues, including the CNS,[40,41] but contain different intracellular domains associated with distinct cytoplasmic proteins and thus mediate distinct cellular responses in vivo.[42] At least in the periphery, p55 TNFR monopolizes TNF-mediated signaling and can signal almost all reported TNF activities including cytotoxicity,[43] apoptosis[44] and proliferation.[45] In contrast, p75 TNFR signaling is limited to a few cell systems and promotes the proliferation of thymocytes, T lymphocytes and other cells of hemopoietic origin,[46] as well as apoptotic signals in mature activated CD8[+] T cells.[47] Recent evidence indicates that the 26 kDa transmembrane TNF molecule is superior to soluble TNF in activating the p75 receptor in various systems such as T cell activation and thymocyte proliferation,[48] and implies an important physiological role for this receptor in local inflammatory responses. In the CNS such a hypothesis has been supported by the work of Lucas et al,[49] who have shown that in contrast to p55 TNFR null mice, mice genetically deficient for the p75 TNFR are resistant to cerebral malaria. In our laboratory we have used transgenic and gene knock-out technology to address specific questions concerning TNF action in the CNS, the differential action of soluble and transmembrane TNF, and the two TNF receptors in CNS immune response and pathophysiology. Transgenic mice have been generated in which wild-type murine, and wild-type or transmembrane human TNF transgenes are expressed by major cell types of the CNS, specifically astrocytes and neurons (Table7.1).[50,51]

Demyelination and Axonal Damage

TNF has been strongly implicated in the pathogenesis of MS,[5] and in particular in the process of demyelination. Demyelination is a hallmark of MS and is thought to be responsible for neuronal dysfunction experienced by MS patients. It involves the destruction of the myelin sheath that insulates the axons of the neuronal cells by aberrant immune mechanisms involving B and T cells, macrophages, complement and cytokines.[52,53] TNF is produced by astrocytes and microglia/macrophages in MS brain lesions[10-12] and TNF levels in the cerebrospinal fluid reflect disease activity in MS.[54] Furthermore, at the perimeter of acute MS lesions myelin vacuolation is observed, which is similar to the pattern of myelin breakdown observed when organotypic CNS cultures are exposed to TNF.[55] In vitro experiments have also revealed that TNF, as well as lymphotoxin-α (LT-α), induce programmed cell death in oligodendrocytes,[13-14] the cells responsible for myelin production which are lost in MS lesions,[56] and studies which show that transmembrane TNF kills oligodendrocytes more effectively than soluble TNF[57] suggest that cell contact dependent mechanisms may be important in mediating such effects in vivo. Furthermore, in vitro[41,58] and in MS lesions,[55,59] oligodendrocytes are found to express p55 and p75 TNFRs, among other molecules of the apoptotic cascade, such as Fas.[17,18]

Several lines of in vitro and recent in vivo experiments have attempted to define the effects of TNF on neuronal viability and to elucidate the signaling pathways which TNF utilizes to mediate such effects. The effect of TNF on neurons has been difficult to define. In vitro data have shown that TNF can induce both neurotoxic, via the p55 TNFR,[27] and neuroprotective[28,29] effects on neurons. This apparent discrepancy may reflect differences in culture conditions, and particularly whether glial cells are present in the culture. Microglia,

Table 7.1. CNS-expressing TNF transgenic mice

Transgene	Line	Cell source	Phenotype	Reference
muTNFgl	Tg6074	resident CNS cells	seizures, ataxia, paralysis, Life span; 3-6 months	50
GFAP-wt huTNFgl	TfK83*		hydrocephalus paralysis	51
GFAP-tm huTNFgl	TgK21	astrocytes	Life span; 3-4 weeks	
NFL-wt huTNFgl	TgK742		paralysis Life span;	51
NFL-tm huTNFgl	TgK3	neurons	3-6 months none	

*extinct line
GFAP, glial fibrillary acidic protein; gl, 3'UTR and polyA of human β-globin gene; huTNF, human TNF; muTNF, murine TNF;
NFL, neurofilament light chain; tm, transmembrane; wt, wild type

for example, have been shown to release neurotoxins that affect neuronal viability.[60] It should also be kept in mind that neurons represent a heterogeneous cell population and the discrepancy between the in vitro data could reflect differences in the responsiveness of neurons to TNF. Interestingly, recent in vivo experiments have shown that neuronal death induced by ischemic injury is enhanced in the absence of the p55 and p75 TNFRs and point to a neuroprotective role for TNF.[30] This neuroprotective function of TNF is thought to be mediated directly through the induction of NFκB and NFκB-regulated genes in neurons,[29] and may involve the active suppression of TNF-mediated cell death signals as has been shown in other cell types.[61,62] Interestingly, a recent study has associated NFκB with neuronal apoptosis,[63] but a causal relationship between NFκB and neurodegeneration has yet to be shown.[64] Transgenic mice which express TNF in their CNS prove useful tools for analyzing the role of TNF upon neuronal viability in an in vivo system where neurons are not affected by in vitro manipulations.

Inflammation and white matter loss are the major pathologic alterations in Tg6074 transgenic mice which express a murine TNF transgene under the control of its own promoter, specifically by resident cells of the CNS.[50] Tg6074 mice spontaneously develop a neurologic disease with 100% phenotypic penetrance which is manifested clinically by loss of the limb flexion reflex, seizures, ataxia, imbalances and partial paralysis. In this transgenic line the first histopathological changes are pronounced astrocytic gliosis in the circumventricular organs and widespread microglial activation.[50] These histopathological changes progress to develop into lesions that show demyelination at the capsula interna and the cerebellum in the presence of phagocytic macrophages (Fig. 7.1).[50] The myelin loss observed at this stage closely resembles that seen in MS lesions. Late lesions of Tg6074 transgenic mice are characterized by BBB (blood-brain barrier) leakage, perivascular and parenchymal lymphocyte infiltration, loss of myelin, abundant macrophages and pronounced astrocytosis. In addition, functional studies have demonstrated that at the time point of CNS lesion development, Tg6074 mice show behavioral defects that may be linked to neuronal dysfunction.[65]

TNF also has the potential to induce the formation of acute lesions when overexpressed by astrocytes in transgenic mice.[51] In mice of the TgK21 line, where a human transmembrane TNF transgene is directed to express in astrocytes, the prime pathogenic effects induced by TNF are the accumulation of inflammatory cells in the meninges of the spinal cord, endothelial cell activation and BBB damage accompanied by pronounced astrocytic gliosis.[51] At sites of extreme inflammation and BBB breakdown, acute lesions develop which are characterized by loss of myelin, neurons and astrocytes (Figs. 7.2 and 7.3).[51] In TgK21 transgenic mice, chronic expression of TNF by astrocytes, cells which form intimate associations with the BBB through their foot processes and induce BBB properties in CNS endothelial cells in vitro,[68] seems to be critical for endothelial cell activation and BBB damage. This contact favors the establishment of a robust inflammatory component characteristic of acute fulminant lesions of the CNS, such as those seen in acute MS. [66,67]

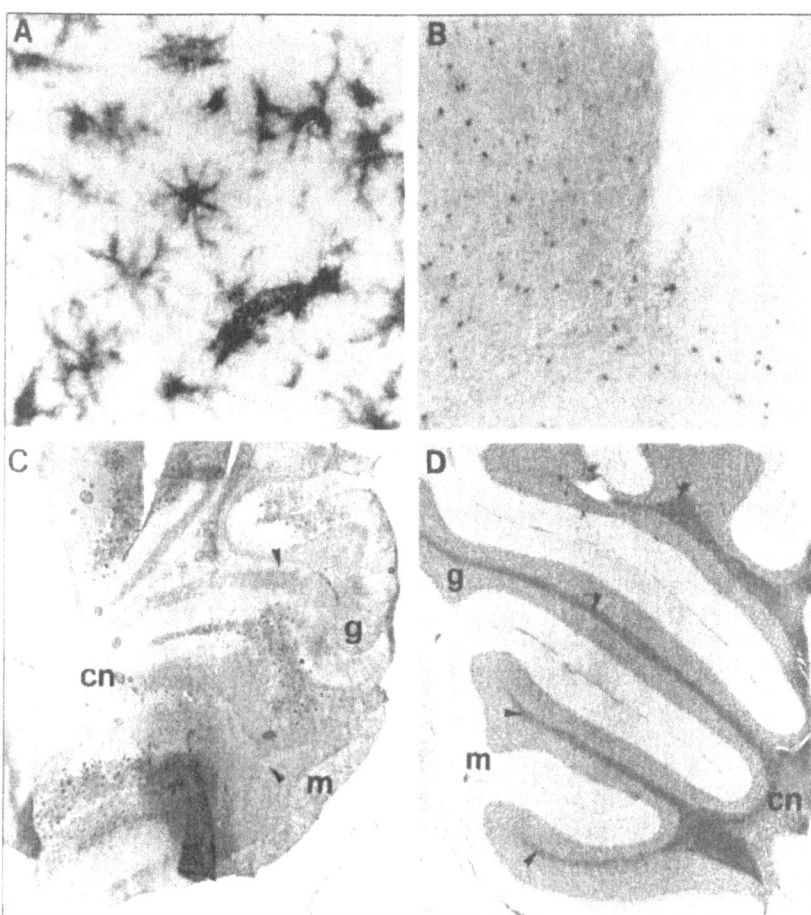

Fig. 7.1. Inflammatory demyelination in the brain of Tg6074 mice. (A) Activated microglia stained with anti-CD18 antibody. (B) Numerous CD8 positive T cells in the CNS parenchyma. (C) Cerebellum of a Tg6074 mouse stained with oil red O and hematoxylin (H) shows loss of myelin (arrowheads) and numerous oil red O positive macrophages. Normal cerebellum stained with oil red O/hematoxylin (D) shows normal distribution of myelin (arrowheads) and absence of phagocytic activity. cn: cerebellar nucleus, g: granular layer, m: molecular layer.

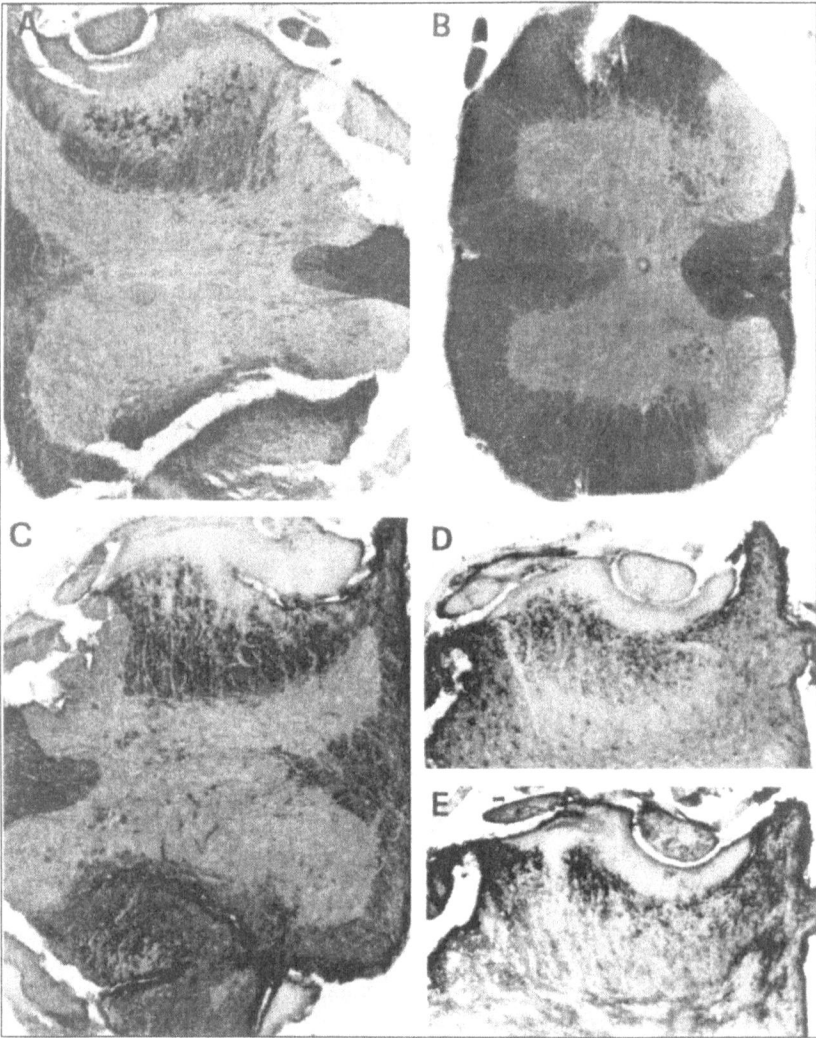

Fig. 7.2. Inflammatory demyelination in TgK21 mice. Oil red O/hematoxylin-stained spinal cord cross section of TgK21 (A) reveals excessive accumulation of lipid debris (oil red O⁺) at the frontier between myelin and inflammatory cells. (B) Normal spinal cord stained with oil red O/hematoxylin shows normal myelin distribution and absence of inflammatory cells. A serial section to (A) stained with an antibody for MBP (C) shows demyelination and reveals that the lipid debris is MBP positive. Staining of serial sections to (A) with an antibody against GFAP (D) or human TNF (E) shows that the astrocytes produce the human TNF transgene locally at the site of the lesion. For color representation see page 152 in Color Insert.

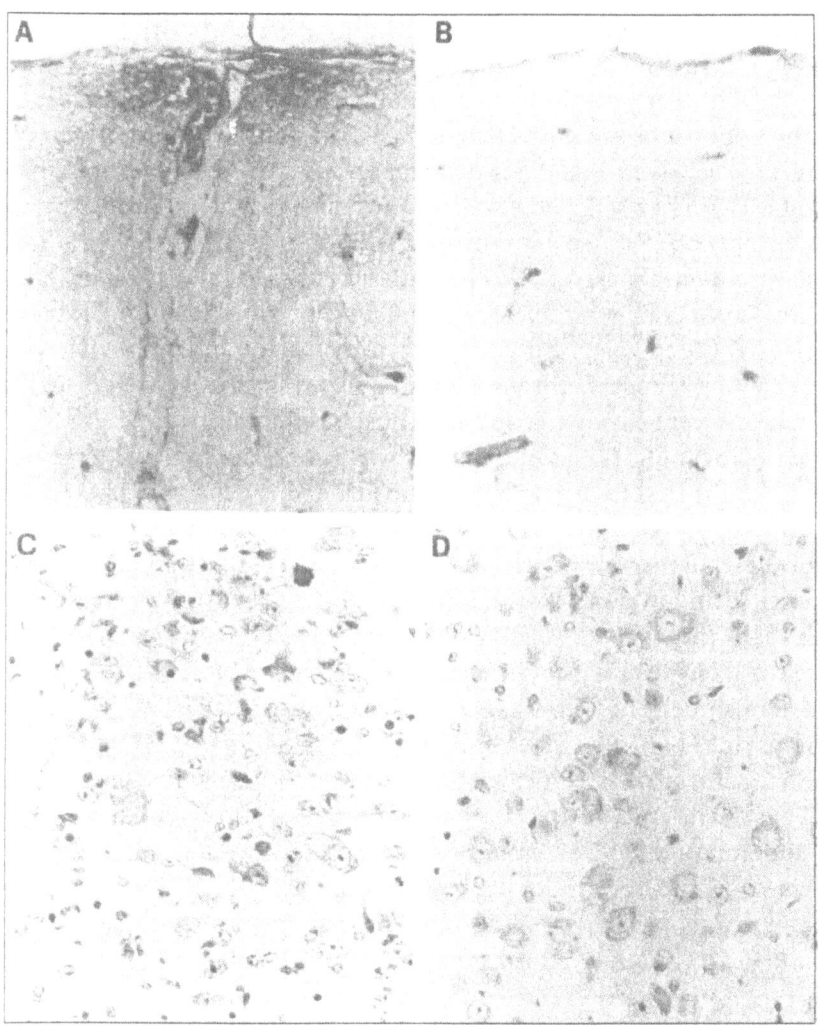

Fig. 7.3. BBB disruption and neurodegeneration in TgK21 mice. (A) TgK21 cortex stains positive with an antibody against serum albumin, showing leakage of the BBB. (B) Cortex of a normal control mouse treated as in (A) is negative for albumin staining in the CNS parenchyma. Luxol Fast Blue/Cresyl violet staining of TgK21 transgenic spinal cords (C) shows degenerative neurons, while neurons of a TgK3 mouse (D) appear normal.

Cellular Interactions and Molecular Mechanisms in TNF Induced CNS Pathology

To directly address the question of which cellular and molecular interactions are important in mediating TNF-induced pathology in the CNS, we have generated transgenic mice which express human TNF as a transmembrane or soluble molecule by astrocytes and neurons, cells which have diverse function, placing and immunologic potential in the CNS. The promoter regions of the murine glial fibrillary acidic protein (GFAP)[69,70] and neurofilament light chain (NFL)[71,72] genes, were used to target expression of either wild type or mutant transmembrane TNF transgenes in mouse astrocytes or neurons respectively (Table 7.1).[51] The mutant uncleavable transmembrane form of TNF was prepared by deletion of the triplets encoding for the first to the twelfth amino acids of the mature 17 kDa TNF protein, which has previously been shown to produce a transmembrane TNF protein which remains bioactive against cellular targets in vitro and in vivo.[73] Microinjection of these constructs into fertilized murine (CBA X C57BL/6) oocytes resulted in the generation of three GFAP-wtTNF founder mice (TgK15, TgK18, TgK83) and the establishment of NFL-wtTNF (TgK742), GFAP-tmTNF (TgK21) and NFL-tmTNF (TgK1, TgK3 and TgK11) transgenic lines[51] (Table 7.1).

All transgenic mice engineered to produce wild-type human TNF in either astrocytes or neurons spontaneously develop a neurological phenotype manifested by tremors, ataxia, seizures and paralysis, and bear histological evidence of chronic CNS inflammation and degeneration. Based on the reported species specificity of the murine p75 TNFR,[74] these data show that the observed pathogenic potential of human TNF transgenes can be signaled exclusively through the p55 TNFR. The observation that both neuron and astrocyte-derived TNF are able to trigger the development of CNS inflammation suggests that, when it is expressed at high levels, soluble TNF is able to act at distant cellular targets within the CNS to mediate endothelial cell activation, perturbation of the BBB and induction of inflammatory infiltration.

In contrast to wild type TNF, which can trigger neurologic disease irrespective of the cell type producing it in the CNS, the transmembrane TNF molecule is able to induce disease when it is expressed by astrocytes, but not neurons.[51] As described above, transgenic mice that express transmembrane TNF by astrocytes (TgK21) develop neurologic disease and show histological evidence of inflammatory demyelinating and neurodegenerative lesions, while expression-matched transgenic mice that express transmembrane TNF by neurons (TgK3) show no phenotypic or histological abnormality. Therefore, in contrast to soluble TNF which appears to exert its pathogenic effects in the CNS independently of the cell-type which produces it, transmembrane TNF acting in a contact-dependent manner needs to be produced by cell types of the CNS that form the appropriate contacts with the cellular targets of TNF in the CNS. This finding demonstrates the pathogenic potency of contact-dependent, astrocyte-specific TNF signaling within the CNS and indicates that endothelial cells and oligodendrocytes, which form intimate associations with the astrocyte, may represent cellular targets of TNF action in the CNS in vivo.

TNF Transgenic Mice as Models for Human CNS Diseases

There is considerable evidence that TNF plays a fundamental role in the pathogenesis of inflammatory autoimmune diseases, including those affecting the CNS. In addition to its association with MS and other human inflammatory CNS diseases, inhibitors of TNF have been used successfully to treat experimental diseases like experimental autoimmune encephalomyelitis (EAE),[75-77] experimental autoimmune uveitis[78] and cerebral malaria.[79] Moreover, susceptibility to EAE correlates to higher TNF production by glia[80] and the encephalitogenicity of myelin basic protein-specific T cell clones correlates with their TNF production.[81] Although recent evidence shows that TNF and lymphotoxin α are not essential for the development of EAE,[82,83] and that EAE may even be exacerbated in the absence of TNF,[84,85] indicating additional immunomodulatory roles for TNF during autoimmune disease, the weight of evidence implies a fundamental role for chronic local TNF production in active CNS disease pathogenesis. Transgenic mice that express TNF in their CNS establish TNF as a potentially major etiopathogenic component of the immune response that is responsible for triggering neuropathologic alterations in the CNS, and the use of human TNF transgenes demonstrates the dominant role of the p55 TNFR in signaling this TNF-mediated pathology. In addition, our results imply that activated astrocytes, a rich natural source of TNF in the CNS, have the potential to induce inflammation, demyelination and neurodegeneration. TNF transgenic mice represent new animal models for the study of neuroinflammatory diseases which are based upon a fundamental disbalance of the immune environment of the CNS, and provide important opportunities for analyzing basic mechanisms which underlie complex human diseases such as MS.[86,87]

Acknowledgments

This work was supported by European Commission Grants BIO4-CT96-0174 and BIO4-CT96-0077.

References

1. Sternberg EM. Emotions and disease: From balance of humors to balance of molecules. Nature Med 1997; 3:264-267.
2. Payan DG. Neuroimmunology. Adv Immunol 1986; 39:299-323.
3. Hopkins SJ, Rothwell NJ. Cytokines and the nervous system I: Expression and recognition. Trends Neurosci 1995;18:83-88.
4. Merrill JE, Benveniste EN. Cytokines in inflammatory brain lesions: Helpful and harmful. Trends Neurosci 1996; 19:331-338.
5. Raine CS. Multiple sclerosis: TNF revisited, with promise. Nature Med 1995; 1:211-214.
6. Meda L, Cassatella MA, Szendrel GI et al. Activation of microglia cells by â-amyloid protein and interferon-β. Nature 1995; 374:647-650.
7. Leist TP, Frei K, Kam-Hansen S et al. Tumor necrosis factor α in cerebrospinal fluid during bacterial but not viral meningitis. J Exp Med 1988; 167:1743-1748.
8. Grau GE, Piguet PF, Vassalli P et al. Tumor necrosis factor and other cytokines in cerebral malaria: Experimental and clinical data. Immunol Rev 1989; 189:49-70.
9. Tyor WR, Glass JD, Griffin JW et al. Cytokine expression in the brain during the acquired immunodeficiency syndrome. Ann Neurol 1992; 31:349-360.
10. Hofman FM, Hinton DR, Johnson K et al. Tumor necrosis factor identified in multiple sclerosis brain. J Exp Med 1989; 170: 607-612.
11. Selmaj KW, Raine CS, Cannella B et al. Identification of lymphotoxin and tumor necrosis factor in multiple sclerosis lesions. J Clin Invest 1991; 87:949-954.
12. Canella B, Raine CS. The adhesion molecule and cytokine profile of multiple sclerosis lesions. Ann Neurol 1995; 37:424-435.
13. Selmaj K, Raine CS. Tumor necrosis factor mediates myelin and oligodendrocyte damage in vitro. Ann Neurol 1988; 23:339-346.
14. Selmaj K, Raine CS, Farooq M et al. Cytokine cytotoxicity against oligodendrocytes. Apoptosis induced by lymphotoxin. J Immunol 1991; 147:1522-1529.
15. Selmaj KW, Farooq M, Norton WT et al. Proliferation of astrocytes in vitro in response to cytokines. A primary role for tumor necrosis factor. J Immunol 1990; 144:129-135.
16. Merrill J. Effects of interleukin-1 and tumor necrosis factor-α on astrocytes, microglia, oligodendrocytes, and glial precursors in vitro. Dev Neurosci 1991. 13:130-137.

17. D'Souza SD, Bonetti B, Balasingam V et al. Multiple sclerosis: Fas signaling in oligodendrocyte cell death. J Exp Med 1996; 184: 2361-2370.

18. Dowling P, Shang G, Raval S et al. Involvement of the CD95 (APO-1/Fas) receptor/ligand system in multiple sclerosis brain. J Exp Med 1996; 184:1513-1518.

19. Pober JS. Cytokine-mediated activation of vascular endothelium. Physiology and pathology. Am J Pathol 1988; 133:426-433.

20. Seelentag WK, Mermod JJ, Montesano R et al. Additive effects of interleukin 1 and tumor necrosis factor-alpha on the accummulation of the three granulocyte and macrophage colony-stimulating factor mRNAs in human endothelial cells. EMBO J 1987; 6:2261-2265.

21. Hurwitz AA, Lyman WD, Guida MP et al. Tumor necrosis factor α induces adhesion molecule expression on human fetal astrocytes. J Exp Med 1992; 176:1631-1636.

22. Lavi E, Suzumura A, Murasko DM et al. Tumor necrosis factor induces expression of MHC class I antigen on mouse astrocytes. J Neuroimmunol 1988; 18:245-253.

23. Vidovic M, Sparacio SM, Elovitz M et al. Induction and regulation of class II major histocompatibility complex mRNA expression in astrocytes by interferon-β and tumor necrosis factor-α. J Neuroimmunol 1990; 30:189-200.

24. Nikcevich KM, Gordon KB, Tan L et al. INF-α-activated primary murine astrocytes express B7 costimulatory molecules and prime naïve antigen-specific T cells. J Immunol 1997; 158: 614-621.

25. Williams KC, Dooley NP, Ulvestad E et al. Antigen presentation by human fetal astrocytes with the cooperative effect of microglia or the microglia-derived cytokine IL-1. J Neurosci 1995; 15:1869-1878.

26. Shrikant P, Benveniste EN. The central nervous system as an immunocompetent organ. Role of glial cells in antigen presentation. J Immunol 1996; 157: 1819-1822.

27. Sipe KJ, Srisawasdi D, Dantzer R et al. An endogenous 55 kDa TNF receptor mediates cell death in a neural cell line. Brain Res Mol Brain Res 1996; 38:222-232.

28. Cheng B, Christakos S, Mattson MP. Tumor necrosis factors protect neurons against metabolic-excitotoxic insults and promote maintenance of calcium homeostasis. Neuron 1994; 12:139-153.

29. Barger SW, Horster D, Furukawa K et al. Tumor necrosis factors α and â protect neurons against amyloid â-peptide toxicity: Evidence for involvment of a κB-binding factor and attenuation of peroxide and Ca2+ accumulation. Proc Natl Acad Sci USA 1995; 92:9328-9332.

30. Bruce AJ, Boling W, Kindy MS et al. Altered neuronal and microglial responses to excitotoxic and ischemic brain injury in mice lacking TNF receptors. Nature Med 1996; 2:788-794.

31. Kriegler M, Perez C, Defay K et al. A novel form of TNF/cachectin is a cell surface cytotoxic transmembrane protein: Ramifications for the complex physiology of TNF. Cell 1988; 53:45-53.

32. Perez C, Albert I, DeFay K et al. A non-secretable cell surface mutant of tumor necrosis factor (TNF) kills by cell to cell contact. Cell 1990; 63:251-258.

33. Gearing AJH, Beckett M, Christodoulou J et al. Processing of tumor necrosis factor-α precursor by metalloproteinases. Nature 1994; 370:555-557.

34. Black RA, Rauch CT, Kozlosky CJ et al. A metalloproteinase disintergrin that releases tumor-necrosis factor-α from cells. Nature 1997; 385:729-733.

35. Mohler KM, Sleath PM, Fitzner JN et al. Protection against a leathal dose of endotoxin by an inhibitor of tumor necrosis factor processing. Nature 1994; 370:218-220.

36. Birkland TP, Sypek JP, Wyler DJ. Soluble TNF and membrane TNF differ in their ability to activate macrophage antileishmanial defence. J Leukoc Biol 1992; 51:296-299.

37. Aversa G, Punnonen J, De Vries JE. The 26-kD transmembrane form of tumor necrosis factor α on activated CD4+ T cell clones provides a costimulatory signal for human B cell activation. J Exp Med 1993; 177:1575-1585.

38. Lopez-Cepero M, Garcia-Sanz JA, Herbert L et al. Soluble and membrane-bound TNF-α are involved in the cytotoxic activity of B cells from tumor-bearing mice against tumor targets. J Immunol 1994; 152:3333-3341.

39. Solorzano CC, Ksontini R, Pruitt J et al. Involvment of 26-kDa cell-associated TNF-α in experimental hepatitis and exacerbation of liver injury with a matrix metalloproteinase inhibitor. J Immunol 1997; 158:414-419.

40. Kinouchi K, Brown G, Pasternak G et al. Identification and characterization of receptors for tumor necrosis factor-α in the brain. Bioch Bioph Res Comm 1991; 181:1532-1538.

41. Dopp JM, Mackenzie-Graham A, Otero GC et al. Differential expression, cytokine modulation, and specific functions of type-1 and type-2 tumor necrosis factor receptors in rat glia. J Neuroimmunol 1997; 75:104-112.

42. Tartaglia LA, Weber RF, Figari IS et al. The two different receptors for tumor necrosis factor mediate distinct cellular responses. Proc Natl Acad Sci USA 1991; 88:9292-9296.

43. Leist M, Gantner F, Jilg S et al. Activation of the 55 kDa TNF receptor is necessary and sufficient for TNF-induced liver failure, hepatocyte apoptosis and nitrite release. J Immunol 1995; 154:1307-1316.

44. Tartaglia LA, Ayres TM, Wong GH et al. A novel domain within the 55 kd TNF receptor signals cell death. Cell 1993; 74:845-853.

45. Engelmann H, Holtmann H, Brakebusch C et al. Antibodies to a soluble form of a tumor necrosis factor (TNF) receptor have TNF-like activity. J Biol Chem 1990; 265:14497-14504.

46. Tartaglia LA, Goeddel DV, Reynolds C et al. Stimulation of human T-cell proliferation by specific activation of the 75-kDa tumor necrosis factor receptor. J Immunol 1993; 151:4637-4641.

47. Zheng L, Fisher G, Miller RE et al. Induction of apoptosis in mature T cells by tumour necrosis factor. Nature 1995; 377:348-351.

48. Grell M, Douni E, Wajant H et al. The transmembrane form of tumor necrosis factor is the prime activating ligand of the 80 kDa tumor necrosis factor receptor. Cell 1995; 83:793-802.

49. Lucas R, Juillard P, Decoster E et al. Crucial role of tumor necrosis factor (TNF) receptor 2 and membrane-bound TNF in experimental cerebral malaria. Eur J Immunol 1997; 27:1719-1725.

50. Probert L, Akassoglou K, Pasparakis M et al. Spontaneous inflammatory demyelinating disease in transgenic mice showing central nervous system-specific tumor necrosis factor α expression. Proc Natl Acad Sci USA 1995; 92:11294-11298.

51. Akassoglou K, Probert L, Kontogeorgos G et al. Astrocyte, but not neuron-specific, transmembrane TNF triggers inflammation and degeneration in the CNS of transgenic mice. J Immunol 1997; 158:438-445.

52. Martin R, McFarland HF, McFarlin DE. Immunological aspects of demyelinating diseases. Ann Rev Immunol 1992; 10:153-187.

53. Williams KC, Ulvestad E, Hickey WF. Immunology of multiple sclerosis. Clin Neuroscience 1994; 2:229-245.

54. Sharief MK, Hentges R. Association between tumor necrosis factor-α and disease progression in patients with multiple sclerosis. N Engl J Med 1991; 325:467-472.

55. Brosnan CF, Raine CS. Mechanisms of immune injury in multiple sclerosis. Brain Pathol 1996; 6:243-257.

56. Raine CS. The Norton lecture: A review of the oligodendrocyte in the multiple sclerosis lesion. J Neuroimmunol 1997; 77:135-152.

57. Zajicek JP, Wing M, Scolding NJ et al. Interactions between oligodendrocytes and microglia. A major role for complement and tumour necrosis factor in oligodendrocyte adherence and killing. Brain 1992; 115:1611-1631.

58. Tchelingerian JL, Monge M, Le Saux F et al. Differential oligodendroglial expression of the tumor necrosis factor receptors in vivo and in vitro. J Neurochem 1995; 65:2377-2380.

59. Bonetti B, Raine CS. Multiple sclerosis: Oligodendrocytes display cell death-related molecules in situ but do not undergo apoptosis. Ann Neurol 1997; 42:74-84.

60. Giulian D, Vaca K, Corpuz M. Brain glia release factors with opposing actions upon neuronal survival. J Neuroscie 1993; 13:29-37.

61. Beg AA, Baltimore D. An essential role for NF-κB in preventing TNF-α-induced cell death. Science 1996; 274:782-784.

62. Van Antwerp DJ, Martin SJ, Kafri T et al. Suppression of TNF-α-induced apoptosis by NF-κB. Science 1996; 274:787-789.

63. Grilli M, Pizzi M, Memo M et al. Neuroprotection by aspirin and sodium salicylate through blockade of NF-κB activation. Science 1996; 274:1383-1385.

64. Lipton SA. Janus faces of NF-κB: Neurodestruction versus neuroprotection. Nature Med 1997; 20-22.

65. Fiore M, Probert L, Kollias G et al. Neuro-behavioral alterations in developing transgenic mice expressing TNF-α in the brain. Brain Behav Immun 1996; 10:126.
66. Raine CS. Demyelinating diseases. In: RL Davies, DM Robertson, eds. Textbook of Neuropathology. Baltimore, Maryland: Williams and Wilkins, 1997:627-714.
67. Ferguson B, Matyszak MK, Esiri MM et al. Axonal damage in acute multiple sclerosis lesions. Brain 1997; 120:393-399.
68. Janzer RC, Raff MC. Astrocytes induce blood-brain barrier properties in endothelial cells. Nature 1987; 325:253-257.
69. Brenner M. Structure and transcriptional regulation of the GFAP gene. Brain Path 1994; 4:245-257.
70. Sarkar S, Cowan N. Intragenic sequences affect the expression of the gene encoding glial fibrillary acidic protein. J Neurochem 1991; 57:675-684.
71. Monteiro MJ, Hoffman PN, Gearhart JD et al. Expression of NF-L in both neuronal and nonneuronal cells of transgenic mice: Increased neurofilament density in axons without affecting caliber. J Cell Biol 1990; 111:1543-1557.
72. Ivanov TR, Brown IR. Interaction of multiple nuclear proteins with the promoter region of the mouse 68 kDa neurofilament gene. J Neurosci Res 1992; 32:149-158.
73. Georgopoulos S, Plows D, Kollias G. Transmembrane TNF is sufficient to induce localised tissue toxicity and chronic inflammatory arthritis in transgenic mice. J Inflamm 1996; 46:86-97.
74. Lewis M, Tartaglia LA, Lee A et al. Cloning and expression of cDNAs for two distinct murine tumor necrosis factor receptors demonstrate one receptor is species specific. Proc Natl Acad Sci USA 1991; 88:2830-2834.
75. Ruddle NH, Bergman CM, McGrath KM et al. An antibody to lymphotoxin and tumor necrosis factor prevents transfer of experimental allergic encephalomyelitis. J Exp Med 1990; 172:1193-1200.
76. Baker D, Butler D, Scallon BJ et al. Control of established experimental allergic encephalomyelitis by inhibition of tumor necrosis factor (TNF) activity within the central nervous system using monoclonal antibodies and TNF receptor-immunoglobulin fusion proteins. Eur J Immunol 1994; 24:2040-2048.
77. Selmaj K, Papierz W, Glabinski A et al. Prevention of chronic relapsing experimental autoimmune encephalomyelitis by soluble tumor necrosis factor receptor I. J Neuroimmunol 1995; 56:135-141.
78. Sartani G, Silver PB, Rizzo LV et al. Anti-tumor necrosis factor alpha therapy suppresses the induction of experimental autoimmune uveoretinitis in mice by inhibiting antigen priming. Invest Ophthalmol Vis Sci 1996; 37:2211-2218.
79. Grau GE, Piguet P-F, Vassali P et al. Tumor-necrosis factor and other cytokines in cerebral malaria: Experimental and clinical data. Immunol Rev 1989; 112:49-70.

80. Chung IY, Norris JG, Benveniste EN. Differential tumor necrosis factor alpha expression by astrocytes from experimental allergic encephalomyelitis-susceptible and resistant rat strains. J Exp Med 1991; 173:801-811.
81. Powell MB, Mitchell D, Lederman J et al. Lymphotoxin and tumor necrosis factor-alpha production by myelin basic protein-specific T cell clones correlates with encephalitogenicity. Int Immunol 1990; 6:539-544.
82. Frei K, Eugster H-P, Bopst M et al. Tumor necrosis factor α and lymphotoxin α are not required for induction of acute experimental autoimmune encephalomyelitis. J Exp Med 1997; 185: 2177-2182.
83. Steinman L. Some misconceptions about understanding autoimmunity through experiments with knockouts. J Exp Med 1997; 185: 2039-2041.
84. Probert L, Selmaj K. TNF and related molecules: Trends in neuroscience and clinical applications. J Neuroimmunol 1997; 72:113-117.
85. Liu J, Marino MW, Wong G et al. TNF is a potenet anti-inflammatory cytokine in autoimmune-mediated demyelination. Nat Med 1998; 4:78-83.
86. Lassmann H, Vass K. Are current immunological concepts of multiple sclerosis reflected by the immunopathology of its lesions? Springer Semin Immunopathol 1995; 17:77-87.
87. Lucchinetti CF, Bruck W, Rodriguez M et al. Distinct patterns of multiple sclerosis pathology indicates heterogeneity in pathogenesis. Brain Pathol 1996; 6:259-274.

Color Figures

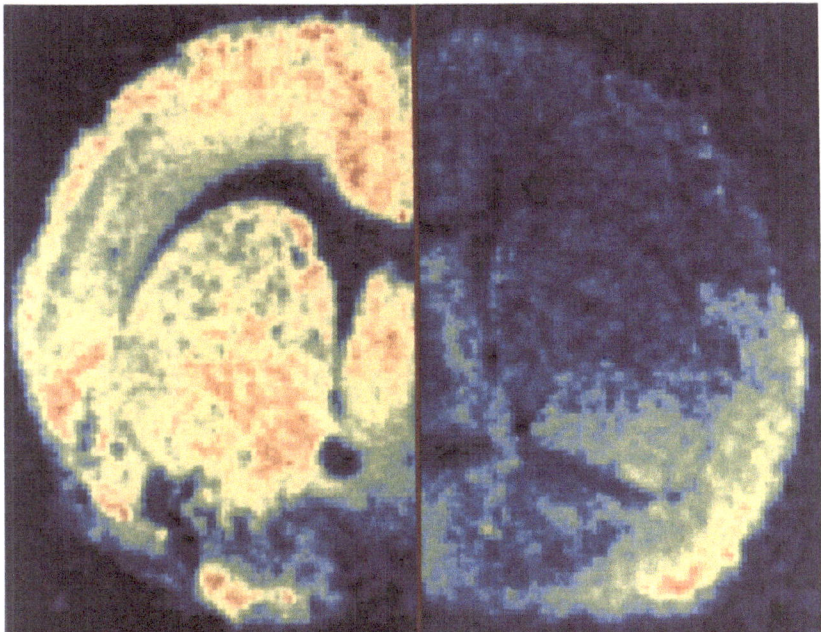

Fig. 5.5. False color coded autoradiograph of [³H]AMPA (30 nM) binding to coronal sections of brain from C57BL/6 mice infected with the LP-BM5 virus for 16 weeks (right panel), and normal C57BL/6 mice. High binding density is indicated by red, low binding density by blue. The density of [³H]AMPA binding was converted from grey scale values to nCi/mg tissue for each section using ³H microscales. [³H]AMPA binding to the somatosensory/motor cortex, lateral septal nucleus, and caudate/putamen from the LP-BM5 infected mouse was 85, 59, and 60% below control values, respectively. Binding density to the piriform cortex was unchanged.

Neuroimmunodegeneration, edited by Paul K.Y. Wong and William S. Lynn.
© 1998 Springer-Verlag and R.G. Landes Company.

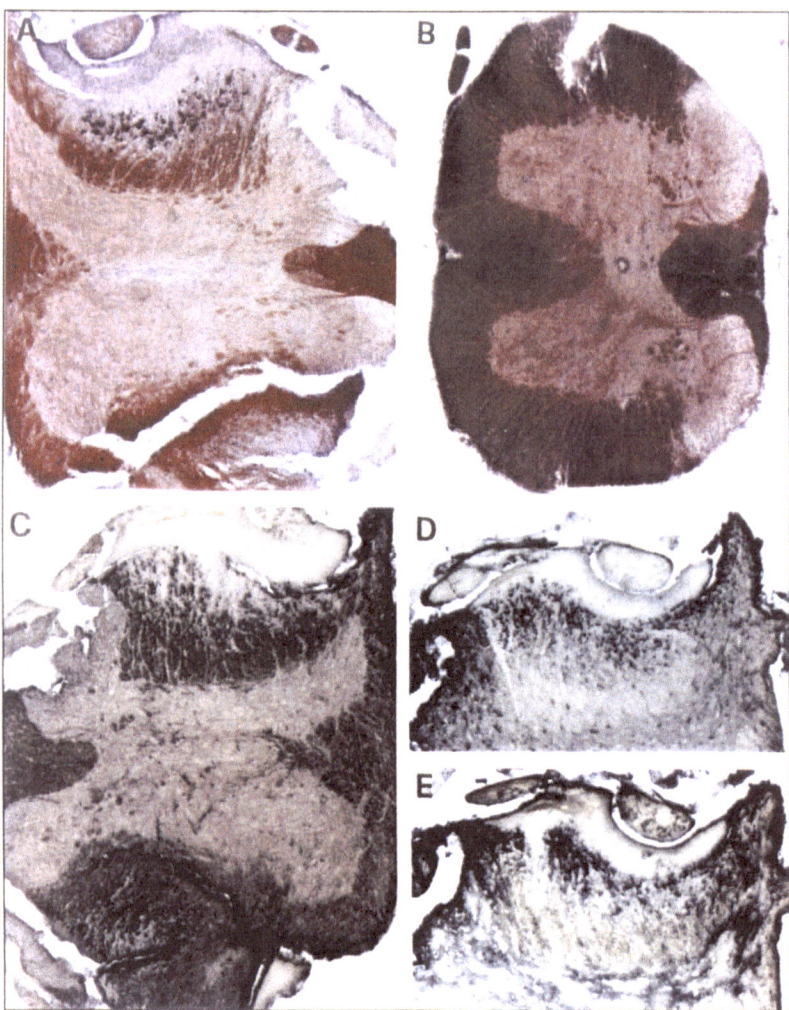

Fig. 7.2. Inflammatory demyelination in TgK21 mice. Oil red O/hematoxylin-stained spinal cord cross section of TgK21 (A) reveals excessive accumulation of lipid debris (oil red O$^+$) at the frontier between myelin and inflammatory cells. (B) Normal spinal cord stained with oil red O/hematoxylin shows normal myelin distribution and absence of inflammatory cells. A serial section to (A) stained with an antibody for MBP (C) shows demyelination and reveals that the lipid debris is MBP positive. Staining of serial sections to (A) with an antibody against GFAP (D) or human TNF (E) shows that the astrocytes produce the human TNF transgene locally at the site of the lesion.

Index

A

AIDS 30, 31, 33, 44, 95, 97, 136
AIDS dementia 31, 33, 44, 95, 136
AIDS dementia complex 31, 33, 44, 95, 136
Alzheimer's disease 30, 31, 37, 116, 120,
124, 129, 136
AMPA receptor 105, 107
Animal models 30, 31, 33, 108, 131, 145
Apoptosis 3, 4, 13-15, 20, 33, 34, 36, 53-57,
61, 62, 64, 65, 67, 79, 88, 96, 125, 137,
139
Astrocytes 4, 6, 8, 18, 31-35, 38-44, 52, 54,
56-58, 61-68, 76, 80, 81, 83-90, 96, 102,
104, 110, 116, 118-120, 124, 125, 127, 128,
136, 138-140, 142, 144, 145
Ataxia telangiectasia 31, 34, 47
Axodendritic process outgrowth 6, 10, 16

B

Behavior 11, 33, 58, 98, 110, 117, 122, 140
Blood-brain barrier 29, 37, 67, 68, 96, 107,
123, 140

C

Caspases 58-60, 62, 67
Cathepsin D 55, 58, 62, 68, 89
Cytokines 1, 2, 5, 6, 8-11, 13, 15, 16, 18, 20,
29, 31-33, 37, 41, 61, 63, 66-68, 79, 82,
84, 97, 102, 115, 116-120, 125, 128-131,
136, 138

D

Dorsoventral patterning 1, 3-5
Down syndrome 30, 31, 40

E

EAE 122, 124, 126, 145
Envelope protein 41, 43, 53-55, 76, 79, 80,
84, 85, 88
ER overload 88

G

Glial development 8
Glutamate 85, 87, 88, 106-108, 110, 111

H

Hemopoietins 1, 10, 14, 15

I

Immunodeficiency 30, 34, 95, 98, 110

L

Lineage commitment 3, 4, 9, 11, 20

M

Multiple sclerosis 116, 120, 124, 129, 136
Murine model 68

N

Neuronal differentiation 6, 9-12, 14
Neurulation 1, 3, 4, 20
Nitric oxide synthase 105, 107

P

P53 35, 59, 60, 64-67
P55 TNF receptor 137
Parkinson's disease 30, 31, 40

R

Retrovirus 30-32, 34, 37, 43, 44, 54, 59, 60,
64, 66, 67, 75, 76, 95, 106, 110

S

Signal transduction 33, 63
Spongiform degeneration 41, 43, 81-83
Striatum 19, 102-106, 108

T

T cells 32, 34, 36, 54, 58, 65, 76, 80, 89, 116,
136, 137, 138, 141
Transforming growth factor beta (TGFb)
128
Transporter 54, 55, 68, 86, 108, 110